THE DRUG MAKERS

A STORY FROM INSIDE THE
PHARMACEUTICAL BUSINESS

DAVID M. SHLAES

ISBN: 978-1-4834-3186-4 (sc)
ISBN: 978-1-4834-3185-7 (e)

Library of Congress Control Number: 2015907896

Lulu Publishing Services rev. date: 6/10/2015

I dedicate this book to the memory of my father,
a brilliant and compassionate physician.

Introduction

This book is about the pharmaceutical industry as seen through the eyes of Daniel Simon. Daniel is a professor of medicine at a Midwestern medical school. He is a specialist in infectious diseases and carries out research on resistance to antibiotics. He moves to the pharmaceutical industry because he wants to make a greater and more direct, practical impact on patient care by finding new antibiotics to treat resistant superbug infections. Daniel works in large companies, small companies, and biotechs, and thus provides a window into all of these different environments.

Large pharmaceutical companies are probably similar to other large organizations in that they are burdened with big, slow bureaucracies that rival small (or even big) governments. They are frequently so slow that they lose any competitive advantage their size might have provided. Constant changes in management and subsequent reorganizations lead to confusion, and depression. It forces managers like Daniel to simply try and keep their heads down.

Personalities play a huge role in the inner workings and ultimate success or failure of all companies, large and small. One thing that Daniel learns is that one great advantage of working in a small company like a biotech is that the CEO is down the hall. One great disadvantage of biotech is that the CEO is down the hall.

The pharmaceutical business is a science-driven business even though pharmaceutical companies spend more money on marketing than they do on research and development. At the end of the day, these companies need to identify products that will address some unmet medical need, will provide some advantage over other competing products and will contribute to the company's bottom line. Daniel's great frustration was that antibiotics were considered by many companies and investors to be unable to provide a reasonable return on investment much less strong profits. All the science in the world would not change that.

Even though this book is only a look at a small and evolving part of the industry – antibiotic research and development – it is meant to provide insight into how the industry functions on a more day-to-day basis.

While one develops a little insight into Daniel as a person, one mainly sees the industry as it unfolds through his eyes, the eyes of someone deeply involved in driving strategic decisions and pursuing his personal goal of getting new antibiotics to patients and physicians.

This book is a work of fiction. Daniel and the characters and companies he encounters are fictitious. The book is also based on the author's experiences over many years and with many companies.

Prologue

Paris, 1995

It was August. Daniel Simon was in Paris. Daniel was in his mid forties with a bushy beard, a graying mustache and chin, and thick but not long hair already graying at the temples. He wore thin tortoise shell framed glasses with round lenses. His career was peaking at his home university in Cleveland. He was nearing the end of a year's sabbatical doing research on antibiotic resistance at the University of Paris. This was Daniel's second such year in the same laboratory in Paris since his first sabbatical there seven years previously. But this time, he was restless and insecure in spite of his apparent success.

The weather was hot and humid. The morning sun blazed through a brown cloud of pollution even though there was little traffic. But the lack of traffic at least made the morning commute by bicycle safe compared to the usual game of dodgem. In August, Paris takes a holiday.

Of course, the bombings that had occurred that summer in Paris made things more interesting. Daniel ran into the results of one on his way home from the lab one evening. He passed a bistro near the St. Michel subway station that had been turned into a makeshift emergency room full of doctors and technicians in white coats and surrounded by emergency vehicles. That bomb killed four people. Daniel and his wife Sally resolved to avoid the subway system for a while, but their resolution quickly dissolved in the face of the practical realities of living in Paris.

Daniel and Sally lived in Cleveland for over 25 years. He went to both medical school and graduate school there earning an MD and a doctorate in microbiology. They married while he was still a student and they raised two daughters. Their daughters were already adults and both were finishing their Masters degrees in business back in Cleveland. Neither had any interest in science or medicine.

Sally was a few years older than Daniel. She was thin and petite, about 5'2" tall, and had short, graying hair dyed brunette. Sally adored simply walking around Paris. She had taken a leave of absence from her job back at the same Veterans Administration hospital where Daniel worked, but had decided that she would retire when they got back from Paris. Sally wanted to spend more time in her garden, with her friends, and she wanted to do some major redecorating of their large home back in Cleveland. Daniel thought that she deserved no less. Sally had, after all, helped pay for his graduate and medical school and had been a constant source of support during their 25 years of marriage.

Back at Cleveland Medical School, Daniel had a moderate sized research laboratory with anywhere from 8 to 15 employees at any one time. They ranged from graduate students to post-doctoral fellows to technicians to nurses all working on various projects around antibiotic resistance. The problem haunting him was – how was he going to continue to fund all their salaries? How was he going to fund his own salary? At CMS he was expected to fund his entire laboratory effort including all salaries, most of his own salary and all equipment and supplies. He had to raise $400,000 per year if not more just to keep his lab running.

For his sabbatical year in Paris, Daniel had applied for 11 different grants from the National Institutes of Health, the pharmaceutical industry and various non-profits like the American Heart Association. The Veterans Administration, where Daniel worked in Cleveland, provided six months of salary support (which included any vacation and sick leave) and allowed him to take an additional six months of leave without pay. But six months salary was not enough for Daniel's family to survive a year in Paris where the cost of living was extraordinarily high compared to Cleveland. Hence his push for grants. Plus, his lab in Cleveland required continued support during his absence.

In general, Daniel would have anywhere from 30-50% of his lab support from the Veterans Administration, and the rest from other sources. He had been lucky as he was awarded a Career Development Award from the Veterans Administration in Washington after completing his first sabbatical year. That grant provided 100% of his salary – so he needed no additional salary from the local VA and nothing from the University. It also meant that, if necessary, when pressed to take on extra duties outside of research, Daniel could say "no". But that grant was about to end. More grant deadlines were looming during the year following his return.

When Daniel tried to examine his situation rationally, his fear seemed unfounded. He had received a large number of grants over the years allowing him to establish an active research laboratory. He published five to ten manuscripts in infectious diseases and biochemistry journals every year and he gave lectures around the world. There was no particular reason why this should not continue to be the case. Nevertheless, his anxiety was persistent.

He was unable to talk to anyone about his fear. He could not even confide in Sally. He knew that Sally would think he was deranged and that she would try and reassure him. But he also knew that would not help. Daniel did not think his colleagues would understand. So he kept this fear to himself and tried to go about his daily routine without thinking about his future too much.

That plan didn't work too well. He started sleeping poorly. "What's going on with you?" Sally would ask.

"Oh, its just the end of the year rush to get manuscripts out. Then there is the move back to the States coming up. I'll be OK."

Daniel Simon was also worried because he had so loved his year in Paris. Both he and Sally loved living in Paris. Neither considered themselves "city people." In Cleveland, they lived in a wooded suburb about 2 miles from the VA Medical Center in Cleveland where they both worked. They had a five-minute commute to work. But Paris was special. Even though living there felt like the hustle and bustle of the city with crowded rides on the subway system (Metro) every day and difficult driving through incredible traffic even at night, Paris got you. Its soaring symbols like the Eiffel Tower and the Sacre Coeur church looked down at you at every turn. Its elegant boulevards and ancient buildings were seductive. Daniel and Sally both learned French. Sally took courses at the Alliance Française while Daniel worked in the lab. They developed close friends in Paris both among workmates and outside of work. They were able to travel outside of Paris every other weekend. It was going to be hard going back home, especially for Daniel.

Daniel was able to carry out his research in Paris without much interruption by academic responsibilities and he could even do his own experiments with his own two hands. This was something he could almost never do in Cleveland. But aside from this wonderful situation, Daniel loved the kind of research he was doing. His project involved a close collaboration between his lab in Cleveland (mainly himself and his resistant bacterial strains isolated from patients at his hospital), the laboratory in Paris that specialized in the biochemistry of bacteria, and SmithKline Beecham, a large pharmaceutical company who provided the expensive instruments that were required to carry out the experiments. This meant constant discussions, meetings and trips between Paris and various SKB sites in England. Each major collaborator brought a unique scientific knowledge base and a certain way of looking at the problem that alone, no single laboratory or scientist could have duplicated. The contributions of each to the group were key to understanding the results of their research. Without this team approach, the project could never have moved off first base.

This broadly collaborative research is rare (even now) in academia. One is almost never one's collaborators' first priority. Frequently, research collaborators compete for the same pot of grant money. But Daniel's sabbatical project united two disparate and non-competing labs (they even shared grants) and labs in a large pharmaceutical company. As Daniel began to think about returning home, he wondered how he would ever be able to duplicate what he had built during this sabbatical year.

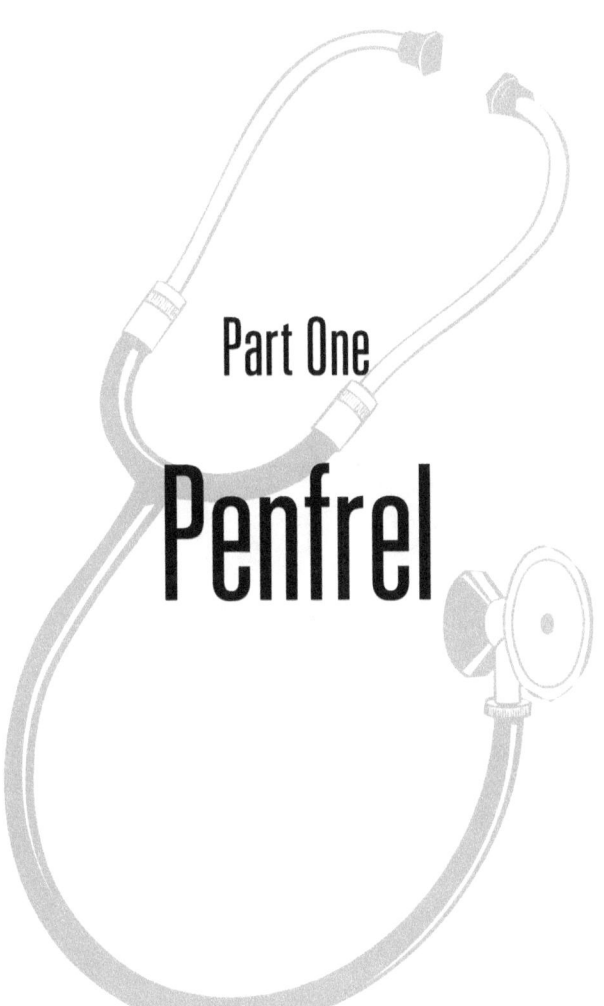

Part One

Penfrel

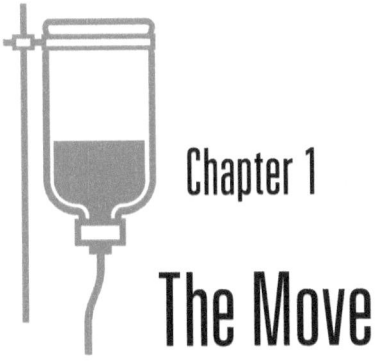

Chapter 1

The Move

Every morning Daniel arrived at work in Paris wondering how he could possibly continue the kind of research he had started in Paris once he returned to Cleveland. Could he come up with the grant money? Could he maintain his collaborations with large pharmaceutical companies? One afternoon, as he was working on writing up results into manuscript form (Daniel and his collaborators would publish six of them in scientific journals), he received a phone call from a headhunter. He occasionally got such calls, usually for academic positions, although once, just after his first sabbatical, he considered the possibility of working in the pharmaceutical industry. He had been interested in the job at the time, but it went to an academic colleague. It turned out that that company was one of the first to abandon antibiotics research altogether firing everyone involved. This time the headhunter was calling regarding a job as vice president of infectious diseases at Penfrel Pharmaceuticals.

"Penfrel?" Daniel asked. "I know the name, but I'm not aware that they market antibiotics."

"Penfrel acquired Lederman Pharmaceuticals in New Jersey over a year ago now. You know them, don't you? Didn't you help them with their major antibiotic product, peracillin?"

"Sure. I know Lederman."

Daniel's interest piqued immediately. It turned out that the physician heading the infectious disease research group at Lederman, now Penfrel, was an old acquaintance, Tom Franklin. Tom was a wonderful infectious diseases clinician and researcher who had taken the reins of the anti-infectives (antibiotics and antivirals) research group at Lederman Pharma. Daniel and Tom had discussed the possibility of Daniel following in Tom's footsteps at some point. And here was the phone call from the headhunter.

"Penfrel is a large company with diverse holdings." The headhunter continued. "Over the years, Penfrel acquired more than twenty smaller companies so that now it has subsidiaries that sell kitchenwares, canned foods, cosmetics for men and women, consumer goods like toothpaste and shampoo, nutritionals for babies, products for animals like drugs and vaccines, and pharmaceuticals and vaccines for people. The company employs about 60,000 people around the world and is very active in European markets as well as in the United States. It brings in about ten billion dollars in revenue every year."

Clearly, Penfrel was a large company with very diverse interests and, more importantly to Daniel, resources.

After several phone calls in rapid succession, the headhunter arranged for Daniel to fly from Paris to Boston to meet the senior vice president for discovery research at Penfrel, Alan Smith. If Daniel were hired, Alan would be his immediate supervisor. As is frequently the case, the company was in a big hurry to fill the position.

Daniel's meeting was for breakfast on a hot, humid August day at the Four Seasons. Daniel arrived the afternoon before. He had to buy a shirt and tie for the interview since he did not have any with him in Paris. He took a cab from his brother-in-law's apartment in Boston, where he had stayed the night, to the Four Seasons. He met Alan for breakfast where they discussed antibiotics, antibiotic resistance, antiviral drugs, management styles, and running. Alan was a serious marathon runner. Daniel routinely ran a puny three to four miles at a fairly slow pace. They took a leisurely walk through the Boston Commons while they continued their discussion.

"So, if you were offered this position, when do you think you could start?" Alan asked.

Daniel had to think for a minute. He would have to pack up and return to Cleveland, get the house ready for sale, resign from the university, - it was overwhelming. "I don't know. Maybe towards the end of October?"

Suddenly, Alan looked distracted and said he had to go to another meeting. The end of their conversation was abrupt. Daniel tried to assure Alan of his interest in the position, because somehow Daniel thought that Alan did not believe Daniel was ready for this move.

Daniel returned to Paris the following evening a little confused, but seriously considering the idea of a move to industry. He had a follow-up phone call with the headhunter that was inconclusive. After that, Daniel heard nothing for months. But the interview got Daniel thinking. If he moved to industry, he knew that he would have three career goals. First, he wanted to make an impact on human health by helping to bring new antibiotics to patients and physicians. The new antibiotics he dreamed about would treat bacterial infections resistant to antibiotics already available. This was based on Daniel's own frustrating experiences trying to treat such patients with life threatening infections where nothing was working. Daniel wanted more of the collaborative research style he'd enjoyed during the previous year in Paris. And he wanted access to the best technology to advance the science quickly. All of these wishes would eventually come true – but not before Cinderella turned into a frog and the carriage into a pumpkin.

In the meantime, Daniel and Sally packed their belongings. He finished various manuscripts and sent them off. They headed back to Cleveland sometime in September. On arrival, with his continuing feelings of unease and insecurity about his path forward in academic research, he began to explore how he could establish or continue the kind of collaborative research he had had in France, back in Cleveland.

Right around the time of Daniel's return, there was a disaster at SmithKline Beecham. Daniel, in addition to collaborating with its scientists, had been a consultant for SKB and had developed close relationships with its antibiotics research group in Brockham Park in

England. Brockham Park was a large estate in the rolling hills of Surrey in southern England. SKB had a famous research site there. It was there that the compound ampicillin was discovered. That in turn led to an explosion in the number of possible synthetic penicillin antibiotics. Just before his return to the United States, Daniel and a team of other consultants had a very interesting and productive meeting with the Brockham Park folks where they explained how they were the first large pharmaceutical company to dive into the new science of genomics to discover new antibiotics. The idea was to sequence the DNA of important bacterial pathogens to identify their genes. These genes coded for proteins and enzymes, and these could then be targeted with new antibiotics. Everyone was enthusiastic.

But, within a week or two of that meeting, Daniel received news that SKB had decided to close the Brockham Park facility. Although a few of the scientists were offered jobs in the United States, most were simply fired on the spot, including Daniel's closest contact there. How could they launch a brand new discovery initiative and at the same time fire most of the people who knew how to discover antibiotics?

It turned out that the SKB decision was prompted by the idea that you could deduce everything necessary to bring a drug to market based on DNA. Genomics would be the be-all-end-all technology. The people making this decision were molecular biologists, gene experts, who knew little about actual drug discovery and development. But somehow they had won the day in their arguments with top SKB management. SKB managers seemed to have forgotten (if they ever knew) that ultimately, any drug one might think one has discovered has to actually be tested against bacteria and against infections in animal models. The very people that SKB had just fired routinely carried out these tasks.

In fact, after arriving back in the US, Daniel received a rather strange call from SKB asking if Daniel would help them with their animal studies of antibiotics.

"What happened to your own veterinary group – the group that did all of your animal testing in Brockham Park?" Daniel asked incredulously.

"None of them were offered positions here in the United States and there is no animal testing group here outside of our toxicology group. And they have no idea how to study antibiotics in animals." He was told.

"What do you want from me, exactly?"

"We wondered if you could take on a number of animal studies for compounds coming from our antibiotic discovery work for a while."

"What's a while?"

"I don't know. However long it takes us to resolve this."

Daniel thought about this for a minute before replying. He immediately understood that if he agreed to do this, the work could overwhelm his own laboratory and would encourage SKB to continue doing nothing to support their own antibiotics research group.

He finally replied, "I can't do this for you. Hire the veterinary group from Brockham Park on some kind of interim basis until you can figure out something else."

With that, the conversation ended. But SKB followed Daniel's advice. The veterinarians from Brockham Park ended up earning more money than those offered full time jobs in the United States. They also had the ability to travel back and forth from their homes in England on a frequent basis – something those who were forced to move to the US could not do.

SKB was eventually forced to rebuild its entire group in order to continue their antibiotic discovery effort. Daniel and many of the SKB scientists, both among those fired and those retained, predicted this – but SKB management apparently did not.

Years later, in retrospect, Daniel realized that he should probably have paid more attention to SKB's ruthlessness as he considered going to industry, but he put this down to an SKB-specific problem. At least he learned that a job in the pharmaceutical industry would never be safe. At the same time, he feared that his life in academia, even though it might be safe, would never be as fulfilling as he wanted it to be.

On the practical side of things, going to industry would mean that he would give up his VA benefits and the possibility of a VA retirement package. If he did make this move, somehow, he would have to try to assure some sort of financial future for himself and his family.

The closure of Brockham Park and the firing of his main contact there also left Daniel without the possibility to continue the collaboration he had started during his sabbatical year. He found that his only real way forward was to apply for a grant that would pay for the equipment that would be needed. That process would take at least a year and there was little guarantee that it would succeed.

Sometime in October, another headhunter called. This time, it was regarding a job at a small biotech in Cambridge, Massachusetts called, Bacto. Bacto had just been established by a group of investors and some sophisticated geneticists to identify new genes that could be targets for anti-bacterial drugs (antibiotics) or anti-fungal drugs. It was using a form of so-called "functional genomics" where the functions of the genes were explored as possible ways to kill the microorganism while at the same time avoiding toxicity to the host (humans). That is, Bacto examined what the genes actually accomplished in the bacterial cell. Their scientists would discern whether or not the bacteria could live without the gene, and if not, would then look to see if the gene had some function that could be inhibited by a new drug.

Bacto was looking for someone to direct the anti-bacterial part of the effort. They thought that someone like Daniel, who was both an infectious diseases clinician and a scientist working on antibiotic resistance, would be best. He jumped at the chance to visit them. Twice. The second time, Sally and Daniel were shown real estate possibilities. Boston was one of the highest priced real estate markets in the country vying with New York and San Francisco for the top honors. The salary in Boston and the poor value of their home in Cleveland compared to houses in Boston were going to be problems. Instead of job and retirement security, Bacto was offering stock options. If Bacto was successful, Daniel would do very well and would not have to worry about retirement. If not – he would worry.

He was very tempted by Bacto's approach scientifically and liked the team they had put together. Daniel felt that they could succeed with their particular approach where others might not since Bacto was emphasizing the functional aspects of the genes it was investigating. That was not the case with many other bacterial genomics companies that were springing up at the time where the only approach taken was to do identify the sequence of the DNA and let someone else figure out the rest of the story.

By December, Daniel knew that Bacto would offer him the job. But Penfrel kept him awake at night. Penfrel was a large company. It could carry out animal testing, which is essential to discovering new antibiotics. Daniel knew that unlike every other therapeutic area, animal models of infection and antibiotic therapy of infection in animals were very predictive of how the drug would work in humans. With the correct animal model and the right experimental design, you could accurately predict how many doses of antibiotic per day would be needed and even how many milligrams or grams of antibiotic you would need for each dose in humans. He thought that having your own capability to carry out animal work would be important to discover and develop new antibiotics. Penfrel also had its own toxicology group, its own manufacturing capacity – Bacto had none of this. At Penfrel, everything could come together. At Bacto, they would actually have to partner with another large company like Penfrel and then would be dependent on the large company to bring whatever drug they might discover together to the marketplace.

Daniel called the Penfrel headhunter. No, no one had taken the position at Penfrel. "Are they still looking?" he asked. Daniel explained his situation and his continuing interest in Penfrel. The headhunter put him back in touch with Alan Smith. Alan had Daniel come out to Princeton, New Jersey where Alan had an office at one of the Penfrel facilities. They had a long discussion.

"So, did you give up on Paris?" Alan asked.

"I don't understand." Daniel looked at him with raised eyebrows. "My sabbatical was always going to end in September. There was never a choice of staying in Paris."

Somehow, Alan had interpreted Daniel's attitude the previous summer in Boston as one that indicated that he would rather stay in Paris than work at Penfrel. Daniel was surprised and perplexed that Alan had gotten that impression. To this day, Alan maintains that that is what Daniel had actually stated. Daniel's memory of their discussion in Boston was an entirely different one.

"Well, I'm glad you got back in touch." Alan said. "As you know, we have not filled the position we discussed last summer. I've arranged for you to speak with a number of vice presidents. If things work out and you work here, you'll be interacting with these guys frequently."

The other vice presidents ran the various other areas of Penfrel like neurosciences, cardiovascular sciences, oncology and women's health. Daniel had dinner with Alan and the head of oncology, Robert Stein. Robert worked at Penfrel's research site in Montvale, New Jersey. That was where the infectious disease group also had their labs and offices. Robert would become an important source of support during Daniel's years at Penfrel and even beyond.

On this visit Daniel learned that Penfrel offered a good retirement package, a much better salary than he could hope for at either Bacto or at the VA, and all the benefits and risks of joining industry. He would have to wear a jacket and tie to work every day. He could fly first class. He would have limousines to transport him when needed. It was not going to be like the Cleveland VA.

At some point afterwards, Daniel met a few of the scientists who would work under him in Montvale, New Jersey. Many of them Daniel knew just because the infectious diseases world is such a small one. To convince Sally that a move to Penfrel was a good idea, the headhunter arranged for them to spend a weekend in the area looking at real estate. He made sure that the properties all had a garden since he had already established that Sally was an avid gardener. He also arranged for front row seats to the Metropolitan Opera in New York and a wonderful dinner at a French restaurant. Sally was completely seduced and so was Daniel.

All of this happened during December and January. By sometime in January, Daniel received a formal job offer from Bacto. He had to put them off while he waited for Penfrel, whose bureaucracy had a global reputation for its lack of speed. By early March, he finally had offers from both companies in hand. Without too much delay, Daniel accepted the offer from Penfrel and expressed his regrets to Bacto. But Bacto and Penfrel would, through Daniel, become close working partners for the next five years.

Daniel would start working at Penfrel in late April. He turned in a resignation letter to the VA and to the medical school. It finally hit both Daniel and Sally that they were leaving Cleveland and with it their many close friends and the house they loved. The departure was bittersweet for Daniel, but was mostly bitter for Sally. They spent many tear-filled hours discussing this plan before Daniel sent in his resignation letter.

Sally and Daniel met just after Daniel came to Cleveland to begin graduate training in microbiology and to start medical school. She worked in one of the research laboratories in the microbiology department down the hall from Daniel's lab. On hearing that Daniel and Sally were dating in January, one of the professors remarked that Sally had "the best pins" in the department. Daniel agreed. They fell immediately and deeply in love. By April Daniel had moved into Sally's apartment. There was no choice of apartments since Daniel was living in a basement apartment in the Hough neighborhood of Cleveland – the area that burned during the riots the year before Daniel arrived. The price was right, but he had already been threatened by one of his neighbors – something about bombing his apartment.

In September, a judge in downtown Cleveland married them. Their parents and a few friends and relatives attended. They had a small party at Sally's parents house in the affluent bedroom community of Hudson, Ohio, where Sally grew up.

Ten years later, Daniel had completed all his training and received both a PhD in microbiology and a medical degree, and completed his internship, residency and fellowship in infectious disease. He landed his first full time job as a physician at the Veterans Administration Medical Center and the Cleveland Medical School. This was a full eleven years after he graduated from college and moved to Cleveland. Only then were Sally and Daniel able to afford the house of their dreams in suburban Cleveland – just a few miles from work. Through all this, they somehow managed to stay close, in love and with an intact marriage – unlike so many of their friends and colleagues.

Fifteen years after moving into their dream house, Daniel was confessing all his fears to Sally and trying to explain why he wanted to make this move so badly. But Sally had already divined all that. She understood and supported Daniel's plan. She understood that this was something Daniel had to do and that she could not step in his way. But this realization could not stop the tears. Sally knew that she now had to go through a sort of grieving process and come to terms with a new life. And she was ready and determined to do just that – as difficult as it might be.

As Daniel approached his D-Day, he was realizing that the move to industry was not going to be simple. After all, in spite of all his experience, what did he really know about discovering and developing drugs? On almost every grant application throughout his career, Daniel had written something like, "If we can understand this (whatever the grant was about), we would be able to design new antibiotics against these resistant pathogens." Now, he was finally going to be able to actually do what he said. But he had no clear idea of how this was done.

During Daniel's interview process, Alan Smith explained that Penfrel was considering eliminating the antiviral research group that was then part of the infectious disease group that Daniel would be heading. Alan felt that since Penfrel marketed no antivirals and that since the group had delivered no drugs during all their years at Lederman Pharma prior to the Penfrel take-over, it made no sense to continue the research effort. Daniel was taken aback by his first encounter with large corporate strategy. He countered that he was not going to start a new job where he had to fire half his staff on arrival. If Alan wanted to close down the group, he would have to do so himself before Daniel's arrival. Daniel offered him an alternative scenario. Daniel would evaluate the group's ongoing projects and provide a non-binding recommendation as to whether Penfrel should continue in antiviral research or not based on the science and current state of the projects. Alan agreed to this plan.

About two weeks before his official start date, Daniel showed up in Montvale to meet those in his future team that he had not met before and to have his company-required physical exam. The Penfrel campus in Montvale was in gorgeous green countryside on the New Jersey border with New York State. The campus had been part of Lederman Pharma and had served as a horse farm for manufacture horse sera back in the 1920s and 30s. The campus was dotted with new buildings of red brick. It reminded Daniel of his university days in Wisconsin.

One of the people he met was Sasha Sikorsky, a Russian immigrant who was running the antiviral discovery group and who would be one of Daniel's direct reports. There was a fairly large community of Russian scientists working at Penfrel as it turned out. Daniel and Sasha shook hands and spoke informally for just a few minutes before Daniel was escorted to the physician's office for his physical. This all occurred on a Wednesday.

Sunday, at some early hour in the morning, Daniel received a telephone call from Iain Cavendish. Iain was a Scot with a light brogue and a charming personality. He was also, Daniel would discover later, a serious alcoholic. But this Sunday morning, Iain was calling to inform

Daniel that Sasha had been brutally murdered by his wife and a cousin the night before. The police discovered them as they were dumping body parts into the Passaic River in plastic garbage bags. Apparently Sasha's wife was worried that an upcoming divorce, occasioned by Sasha's extramarital love life, would interfere with the life style to which she had become accustomed. She convinced a cousin to help assure her of an uninterrupted life of luxury without the aggravation of a philandering Sasha.

Sasha was loved and respected by his employees, if not his wife, who were now about to deal with horror and grief after his sudden and brutal demise as well as with the prospect of losing their own jobs. They knew that Penfrel had them in mind for the chopping block. Daniel asked Iain, who was Sasha's second in command, to take over Sasha's duties temporarily until he arrived and could decide how they would proceed. Other than offer sympathies, there was little Daniel could do except to reassure them that their jobs, for now, were safe. They just had to hang in there until they could go through the project evaluation process that Daniel had planned. But Daniel's life for the next several weeks and months was about to be turned upside down as he tried to deal with the grief, insecurity and fear of the team in Montvale.

The day after hearing the grisly news about Sasha, Daniel called Bos Killington. One of the things Daniel was learning even before he started working at Penfrel was that when you are a consultant to a company, you are smart and they listen to you. When you actually work for the company, you need a consultant. Daniel wanted Bos to consult for him. Bos and Daniel had worked together on a number of committees for various professional societies and the National Institutes of Health over the years. Daniel was always working on antibiotics and antibiotic-resistance while Bos worked on antivirals and antiviral resistance. They liked and respected each other even though they did not know each other personally very well. Bos was about five years older than Daniel, but looked older than that. He was almost completely bald, but the hair he had around the edges of his head was either long and gray or short and gray depending on Bos' mood or the season of the year. He had a well-trimmed but almost white full-face beard. Bos was always enthusiastic and was a bit of an optimist. Daniel knew that Bos was not afraid to express his opinions even in the face of opposing strong opinions. Bos had an ego – but he was not overbearing. Bos also had the advantage of knowing everything that was going on at many other companies and in the academic world as well as far as antiviral research was concerned. Like Daniel, he was a physician and a scientist and could provide a clinical perspective to Daniel's researchers beyond Daniel's own view. Daniel knew that he would be out of his depth in trying to review the Penfrel antiviral programs. But he thought that he could get Bos to help. In fact, Bos quickly agreed.

For the next two weeks or so, Iain and Daniel would speak every day or so discussing the state of affairs within the group and planning project reviews. The antiviral research group consisted of about 15 scientists, all virologists. The chemists, the motor of pharmaceutical research programs, were in a separate, central organization within Penfrel. In this way, Alan, the discovery head and Daniel's boss, could control the research flow and balance efforts within the different therapeutic areas. To slow down projects of lower priority, all he had to do was withdraw chemistry support. Projects that were promising and high priority got more chemists and those that were early, risky or not priorities got few or none. Daniel quickly discovered that the antiviral group was running over 10 different projects, all early, but some more advanced than others. He could not understand how this could be accomplished with just 15 virologists. That averaged out to just over one scientist per project.

Daniel moved to a condo in New Jersey close to work in late April of 1996. The trees were in full flower, the daffodils were in bloom and the grass was already green. The sun shone most days. It was Spring in the mid-Atlantic. Life was beautiful. Sally and their dog, a golden retriever named Cassie, stayed home in Cleveland trying to get their house in shape, arrange for a mover and saying goodbye to friends they had made over the previous 27 years in Cleveland. Either Daniel or Sally would commute so they could be together most weekends. They were leaving behind the house Sally thought they would grow old in.

Both their daughters were staying behind, one in Cleveland and the other in Columbus, Ohio. Cleveland Medical School had thrown them a wonderful farewell dinner where Daniel could not help tearing up with emotion. He could barely get through his few words of farewell.

After a few weeks of searching, they found a beautiful house in northern New Jersey just 5 miles from work, in the country. They had a well and septic system for the first time. The house was just 45 minutes by car from New York City and came with a swimming pool, hot tub and an acre of land. It was smaller and also cost over twice what their house in Cleveland was worth. Welcome to the East Coast.

They settled in together in June. They subscribed to the Metropolitan Opera and drove into New York City about once every month for dinner or a show or shopping or all of the above. They had to find a way to do this around Daniel's work schedule, which, at the beginning, was grueling. Daniel was working 12 hour days and weekends trying to understand what his group was doing and how a large pharmaceutical company actually worked. He eventually succeeded at the former, but remained mystified by the latter in many ways.

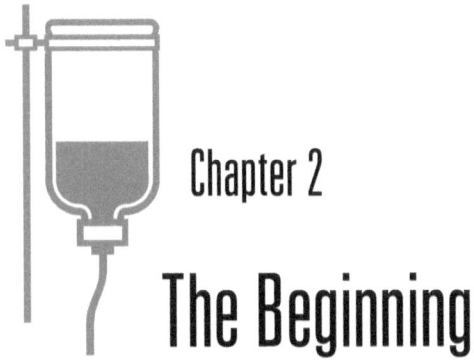

Chapter 2

The Beginning

By the time he arrived at Penfrel, Daniel began to understand the enormity of the job he had taken on. He had a group of about 35 employees. There were two "assistants" (we used to call them secretaries). Both were from New York and had heavy accents. When Sally first called him at work, she could not understand what they were saying. Half of the scientists were PhDs and most were excellent scientists. Many had strongly held opinions especially about their own favorite projects. But Daniel's job was to focus them on those projects that would meet important medical needs stretching 10-15 years in the future, had the greatest chance of success and also had enough commercial potential to interest Penfrel and its shareholders. Of course, Penfrel's idea of commercial potential and Daniel's own beliefs were frequently not in agreement. How do you predict looking into some crystal ball what the situation might be in the 10 to 15 years it would take to get a product from Penfrel's labs onto the marketplace?

During that first six months or so, Alan would meet with Daniel almost every week. He pointed Daniel in the direction of various resources and support systems within Penfrel including scale-up chemistry, manufacturing, drug safety and metabolism and other key departments. His support and counsel was needed and Daniel appreciated it enormously. He could not have asked for more (except when they disagreed as to the need for additional chemists to support a project).

Robert Stein, the head of the oncology research group, had a good deal in common with Daniel. Robert was stocky, gray, balding and looked older than his 55 years. But, like Daniel, Bob had both an MD and a PhD. Before joining industry, he cared for patients and had carried out a research program at MD Anderson Cancer Center in Texas. Bob had been in the industry about 15 years when Daniel arrived at Penfrel. Penfrel was his second company. He had been lured away from his prior company by Penfrel's head of research and development, George Finkel. They had been old friends. Bob had an enormous perspective on the politics of the pharmaceutical industry in general and those of Penfrel in particular. Daniel and Bob had offices just across the campus from each other – they could wave to each other from their windows. Daniel would come to rely heavily on Bob's experience, especially early on when he was climbing his very steep learning curve. Bob had a calm demeanor and almost never lost his temper (at least in front of Daniel). Daniel, on the other hand, could become unnerved and show it. When this happened, he would frequently give Bob a quick call for that soft, calming tone of voice and Bob's perspective that this isn't so bad and that it could always be worse. Bob always said that it made no sense to obsess over problems over which you have no control. This was a hard lesson for Daniel to learn – but oh so valuable.

One of the first things Daniel learned was that all of those support systems within Penfrel (with a few notable exceptions) were headed by people who often thought of their own careers first and everything else later. The irony was that he had taken a job in a large company precisely because they had everything he needed at his fingertips. But, in fact, his needs were only going to be met if they didn't somehow conflict with some department head's idea of the political winds within Penfrel. For the infectious diseases research group, this turned out to be a huge problem because the headwinds were ever present.

Penfrel itself had forgotten its ancient history in antibiotics and infectious diseases. It was one of only a few American companies that helped the British team of Florey, Chain and Heatley manufacture penicillin to test the drug in small clinical trials, then, later, for soldiers fighting in World

War II and finally for the general marketplace. Since then, though, Penfrel had not had an active antibiotic discovery program. They acquired Lederman Pharma in the mid-1990s in order to gain access to new therapeutic areas including antibiotics, vaccines and cancer. But in spite of Penfrel's own history, they didn't really understand these areas of research, sales, marketing or almost anything else about these fields. So department heads were reluctant to put their feet in these strange waters. This all meant that a large part of Daniel's job was going to be dedicated to "teaching" the powers that be within Penfrel that infectious diseases was an important area for research and that it could be profitable as well. He would also have to teach them something about the science behind antibiotics and antivirals, about infectious diseases in general, and about the infectious disease market. This meant that Daniel had to learn a huge amount as well. This seemed crazy since, through their purchase of Lederman Pharma, Penfrel was already marketing several commercially successful antibiotics in hospitals around the world. One of them, peracillin, was predicted to be a billion dollar seller.

Daniel began his new job with reviews of the antiviral projects first since that group seemed to be living on the edge since the murder of their director and the ever-present threat of layoffs from Penfrel. Lederman Pharma had made an early decision that they would not search for drugs for HIV (human immunodeficiency virus), the cause of AIDS. They reasoned that many other companies had already been working in this area for a decade and just had too much of a head start and much more expertise and experience in the area than Lederman and that therefore they could not be competitive. The other reason, Daniel always suspected, is that conservative Penfrel managers did not want to deal with the activist AIDS community of the 1980s and 1990s.

During Daniel's first two weeks, the scientists in the antiviral area presented all the projects they were working on. Most of the projects targeted acute viral disease and not more chronic viral infections like AIDS. This was a potential problem commercially since to treat chronic viral infections, the drugs would have to be taken for years if not for the rest of the patient's lives. Drugs for acute infections would only be taken for a few days to a week or so.

Two of the old Lederman projects were more advanced than the others, but all seemed very academic and not really focused on getting drugs to the patients and physicians who needed them. Also, the Lederman projects were undertaken on a much smaller scale than those in other companies and were scientifically disadvantaged. For example, most large pharmaceutical companies screen very large libraries of compounds to try and find molecules that might work. At the old Lederman Pharma group, small libraries of tens of thousands of compounds had been screened – a fraction of the hundreds of thousands of compounds routinely examined at other large companies. The compounds in the library, Daniel found to his dismay, were largely derived from old dyes and other chemicals that bore no relation to actual drugs that might someday be administered to people. Many were simply so insoluble as to be like rocks or sand.

In their screening, the scientists found compounds that inhibited the growth of the various viruses of interest for a given project. Viruses grow in human or mammalian cells. The problem was that it was frequently hard to separate those compounds that might be specific antivirals from those that might be subtly toxic to the cells in which the virus was growing. Also, frequently, the compounds were so insoluble that subtle precipitation out of solution interfered with the various assays leading the scientists to believe the compound was active when, really, it was just messing up the assay in some subtle way. These problems kept Daniel awake at nights. Daniel was somewhat reassured by the fact that the Penfrel scientists had screened the compounds against several viruses at the same time. If all were inhibited, it suggested that there was some non-specific effect going on – lack of solubility, toxicity to host cells, etc. If only one of the viruses was affected, the compound was likely to be specific for inhibiting that particular virus (a good thing).

Having identified compounds that they thought acted against the virus, the scientists then struggled to understand how these compounds might be working. One sure way was to isolate viruses resistant to the compound. This required a compound that specifically targeted the virus and not its host cell and which was potent enough to allow for selection without other problems. Of the ten or so projects Daniel reviewed, two seemed interesting and might, he thought, actually lead to useful drugs. One targeted Respiratory Syncytial Virus (RSV) – a cause of severe respiratory infection in infants, adults and the elderly. Epidemiologically, RSV caused as much morbidity and mortality each year as the flu! Because there was no therapy, no one bothered to try and diagnose the infection when it did occur. The main reason to diagnose the infection then was to assure the physician that the patient was suffering from a viral infection that did not require antibiotic therapy. The compound identified by the Penfrel scientists, in this case, was used to select a resistant virus successfully. The resistant virus had already been characterized and was shown to have acquired a mutation in the gene encoding the viral protein responsible for fusing the viral membrane with the membrane of the host cell. A number of these mutants had been isolated independently. This assured the scientists, and Daniel, that the compound was acting by interfering with the viral fusion protein and preventing infection of the human cells by blocking the virus' ability to gain entry.

The other interesting project involved a drug for cytomegalovirus (CMV) infection. CMV caused severe infection in immunocompromised patients, especially those with AIDS, cancer patients receiving chemotherapy and those with transplants who received immunosuppressive therapy so they would not reject the transplant. Unfortunately, this therapy made them more susceptible to infection. During March, Daniel's last month at CMS before his move to Penfrel, he was on the infectious diseases consult service for the hospital. During that month he consulted on two patients who had AIDS and had acquired CMV infections of the eye. Both viruses were resistant to the standard anti-CMV drug used at the time. Both were being treated with two alternate drugs, both toxic to the kidneys and neither as effective as the standard drug. Neither patient was responding well to this therapy and their eyesight was threatened. Daniel understood the potential value of a new, non-toxic drug to treat CMV infection. In the case of the Penfrel anti-CMV project, the scientists also had evidence that the compound blocked fusion, but no resistant virus had yet been obtained. This was a cause of some concern for Daniel and the team. Nevertheless, after the two weeks of review, Daniel and his team decided to focus most of the effort within the antiviral group on these two projects and to try and identify new projects with commercial interest for Penfrel. They would eventually go on to initiate research projects on Herpes Zoster Virus, the cause of chickenpox in children and shingles in adults, and on Hepatitis C Virus, the most common infectious cause leading to liver transplantation around the world.

Daniel asked Bos to come up to Montvale to review a few of the projects. The antiviral team was irritated (or at least Daniel thought they were) when they were asked to re-present some of their projects to an outsider they had never met. But Bos quickly won them over and agreed with Daniel that the RSV and CMV projects were the most promising and should be continued if Daniel could get the Penfrel management to agree.

During these two days of meetings, Daniel and Bos developed a close working relationship. One night, Daniel took Sally and Bos out to dinner together. Bos was a connoisseur of port – and even after a fair amount of nice wine at dinner; he always had to finish his meal with a nice (sometimes not cheap) glass of port. Sally found Bos to be endearing. They talked about their families. Bos and his wife Jeanne (she was French) had four children including one who had physical and emotional disabilities but who was able to go to school and was doing well. Daniel and Sally had two grown daughters. From that point on, whenever Daniel would invite Bos to consult at Montvale, and this usually occurred every few months, he would ask Bos to stay at the house with him and Sally. During all the years Daniel and Sally saw Bos regularly, including on occasional trips down to Atlanta (Bos was a professor at Emory University) for one reason or another, they only met Jeanne one time – for a quick dinner out. Daniel and Sally always suspected there was some sort of discord between Bos and Jeanne, but they never knew and never pursued it with Bos or anyone else. Bos' stays at the house worked most times, but not always. Bos traveled even more than Daniel did and seemed to always be on a plane. He sometimes would arrive at Montvale on an early morning flight from Atlanta and he would leave in the evening right after the meetings to head to Washington or San Diego or somewhere else.

As a result of their reviews, Daniel wrote a report to Alan Smith. Daniel spoke of his new consultant Bos and how they had agreed that most projects should be abandoned for various reasons. Daniel recommended, with Bos' agreement, that the projects targeting RSV and CMV continue and that Penfrel maintain their antiviral group with these programs. Alan expressed an interest in meeting Bos during his next visit – something Daniel was anxious to do in any case. Alan agreed with Daniel's plan and then had to try and get his own management, including George Finkel, the head of research and development at Penfrel, to agree as well. Daniel suspected that Penfrel management had already made plans to save money after their acquisition of Lederman Pharma by eliminating the antiviral group. Daniel's recommendation would mean they would have to change their financial plans. In the end, they did as Daniel had recommended.

Abandoning most of their ongoing projects created a good deal of consternation within the antiviral team, but they all recognized the importance of focus and of bringing some critical mass to bear on each problem. They also recognized that their jobs were on the line and that this strategy had saved their rear-ends.

After getting approval to continue antiviral research, Daniel appointed Iain Cavendish to head the antiviral research group.

Chapter 3

Living with Failure

One of the first things Daniel learned on arriving at Penfrel was that many aspects of working in industry were different, as in diametrically opposed, to working in academia. For example, in academia, it's publish or perish. In order to obtain continued grant funding and to be promoted in the university, one is obligated to publish scientific data in respected journals on a very regular basis. When Daniel left the University, he was publishing an average of 5-10 such manuscripts per year.

At Penfrel, the rule was, no publications prior to late stage clinical trials or project failure, whichever should come first. The reason for the rule was to be sure that all the patents required to protect any new compound being studied were filed and to make sure that the competition did not know where Penfrel was at any given time with any given compound on which they had already filed patents.

In academia, a carefully conceived idea leads to a scientifically sound experiment to test the idea. Such projects could last throughout the academic life of a university scientist. Frequently, the result can be published even if you disprove the hypothesis behind your idea. Even if your idea is out of the boundaries of your current research, you can still pursue it. This is called creativity or innovation. In industry, to a certain extent, the scientist is more restricted. It is more difficult for the scientist to pursue his or her own scientific ideas if they fall outside the boundaries of the specific project they have been assigned. Even within the boundaries of the project, experiments must be carefully chosen. Also, given that every scientist is fully burdened with specific projects, it is simply hard to fit in independent work. Many scientists are already working long hours and even weekends in industry, just as they do in academia. Nevertheless, the industry frequently depends on such innovative independent work by its scientists to develop new projects. Prioritizing work for a scientist is a never-ending struggle of competing requirements and a desire and will to innovate.

In academia, you must compete for grant money to support projects and competition is fierce. In industry, an idea for searching for a new antibiotic, for example, has only a certain small chance of actually making it into the official drug discovery process within a company. This is because every such project will cost money and time. Therefore, some level of management must approve the project. In the case of Penfrel, that management was definitely involved. In this case, first, Daniel, but then his colleagues and Alan Smith would have to agree to move the project forward since all would be giving up resources to support this effort. Why? Because even the early stages of the project might involve central resources at the company like using the extensive compound library, the team involved in screening the library using massively automated methods, or using more chemists. Penfrel had an entire team devoted to running tests of the compound library against various therapeutic area targets like bacteria, bacterial enzymes, enzymes from neurosciences, cells or cell receptors for oncology, etc. This team had only so many slots for such screening campaigns every year and each therapeutic area had to compete for these slots.

Within each therapeutic area, there were only so many chemists assigned. So, to move forward with any project, some effort from a chemist would be required. Chemist's time was one of the most valuable resources in the company and was highly prioritized. Again, for the individual scientist to move his/her project forward, the same groups would have to agree. Of course, there were always subterranean projects going on at Penfrel. Somehow Daniel's scientists would embark on an idea and convince a chemist to help by providing compounds with specific characteristics. Sometimes, Daniel's scientists would even surreptitiously obtain compounds from the Penfrel chemical library to look for compounds with some activity against a virus or bacteria. Having found such compounds, the argument to obtain additional resources would be easier.

Once grant funding is acquired in academia, success is the rule. In the pharmaceutical industry, failure is the rule. Even if approval was obtained at Penfrel, and a large screening program was undertaken, the chances of this resulting in a molecule that would make it all the way to the market place is less than one in twenty. Even a molecule that makes it all the way into early testing in humans has only a one in five chance of making it all the way to the marketplace.

These circumstances can be demoralizing for the scientists and this was an important personnel management issue for Daniel and the other managers at Penfrel. Daniel would tell his scientists to focus on the science and, when justified by the data, to write a draft manuscript complete with tables, figures and literature citations. When the time was right, which would occur when the project was officially terminated or when they arrived at phase II stage clinical trials, they could pull out their manuscript, update the draft and submit it. Of course, since they would quickly be assigned other projects as their project failed or moved on to the clinic, they would have to do all the manuscript work "on their own time."

Some of this has evolved in industry – more in some companies than others. In the antibiotics area, you are almost obligated to publish some data showing activity of your compound prior to phase II clinical trials just to help recruit physicians to participate in the trials.

Another difference between academia and industry has to do with collaboration. In academia, you are almost never your collaborator's first priority. Their priority is their own funding status and their own grants. The same is true for you. You might even be competing with your collaborators for the same pool of grant monies.

In industry, once a project gets going, the project is the goal for everyone and is everyone's priority. The farther the project progresses, the more important the group priority becomes. The magnificent part of this is that early in the life of the project, the scientist can bring the most sophisticated techniques and instrumentation to bear on the scientific questions posed. A team can be as large as 50 or more scientists and participants from other disciplines. The variety of scientific domains and technologies that can be focused on the project is astounding and exciting. This can almost never be achieved in academia. At Penfrel, Daniel himself would frequently go to these project team meetings just to see the latest data and to help with strategy on how to move the project forward and which kinds of experiments might be the most helpful to answer the key questions that continually arise. Science is still science – academia or industry. More data leads to more questions. Every answer provides another question and someone still has to decide which question to pursue now and which to leave until later. For Daniel, that was the most fun. That was the thing that kept him coming to work every day. It certainly wasn't filling out forms and budgets for Penfrel management. But, overall, working life for a scientist in industry is very different from that in academia.

Daniel made a few friends among his workmates at Penfrel. These included most especially his boss, Alan Smith, Richard Noland who drove Daniel crazy with his multi-tasking and Bob Stein, his soul mate from oncology across the campus. When Daniel first arrived at Penfrel, the company would provide for a couple of social get-togethers for the various research groups every year – usually around the Christmas holidays. At these parties, Alan and Bob Stein would frequently show up, as would a number of the chemists who worked closely with Daniel's infectious diseases group.

When Penfrel stopped funding the holiday parties for the various departments in favor of one large company-wide party that Daniel and his group never attended, Daniel and Sally began hosting the department Holiday Party at their house in New Jersey every year. This started as a potluck where everyone brought a dish to share – but the preparation and clean up became overwhelming – especially as Daniel's group grew to over 100 employees. With the chemists and spouses and significant others, a couple of hundred people would show up at the house. With Bob Stein's help, Daniel finally found a way to get the party catered and reimbursed by Penfrel. He called it a year-end summary meeting. Sometime during the party, Daniel would call all the employees into the family room of the house and say a few words about the accomplishments of the year. Each department head would make a few comments as well. The entire "meeting" lasted all of five to ten minutes. But Penfrel graciously reimbursed Daniel for the expense of the get-together. The research group, large as it was, functioned like a well-oiled machine with good camaraderie and sometimes, strong personal ties among co-workers.

As Daniel quickly learned, once you start working for a company, you immediately become stupid. After speaking to his new colleagues in marketing and to his scientists, he realized that he would have to establish advisory committees of outside physicians and scientists to vet his decisions and even his leadership. Daniel himself had served on such advisory boards and understood this dynamic well. Through his years in academia and research, Daniel had established a network of friends and colleagues in infectious disease research – both clinical research and more fundamental scientific research.

The relationship between academic scientists investigating the drugs coming out of industry and industry scientists and marketers has always been a delicate one. Companies need the advice of those outside for a variety of reasons. Daniel understood that if you became too insular and isolated, you would never be able to understand medical need – either current or emerging. Since it would take ten to fifteen years for Penfrel to take an idea from the laboratory to the marketplace, the company needed to understand emerging trends in antibiotics and antibiotic resistance to be successful. At the same time, outside criticism of the scientific and clinical work undertaken by Penfrel was important. Without another set of eyes on your research, your own biases could lead you astray. Sometimes an outside view with less invested in a project can alter your approach to a problem.

Academic scientists, in order to maintain their own self esteem and to be of true value to their industry colleagues, were required to be objective in their assessment of company projects – not just for Penfrel, but for all. They are placed in the uncomfortable position of being paid, sometimes well paid, to be critical. Some do better at this than others. Daniel thought long and hard about this conundrum both when he was an academic and again after joining Penfrel. He never came up with a good answer to this. Years later he agreed that making this relationship open and transparent through public disclosure was probably the best one could do in this regard.

Finally, there is the philosophy of "no." Daniel quickly realized that in the pharmaceutical industry, it is always (almost) less costly to say no. It is also more costly to continue a program that is almost certain to fail. These competing tensions are constant in the industry.

The naysayers take almost no risk in saying no. A program that is halted is one where it is very unlikely to know whether it would have ultimately been successful or not. The only time that occurs is when another company wants to license the program or compound that the naysayer has killed. If the other company is successful and makes it to the market with a good selling product, then the naysayer at the original company might be in trouble. But the way things work in the industry, that naysayer has most likely gone on to bigger and better things by the time the product he/she killed hits the market for the competing company. Look at daptomycin. It was discovered at Eli Lilly in the early 1980s, killed by them in the mid 1990s and marketed by Cubist in 2003. Where is the naysayer at Lilly? Daniel had no idea.

Daniel had personal experience with this. At one point his team at Penfrel were interested in licensing in antibiotic from Gorman, Ltd in Pennsylvania. Gorman was very surprised that Penfrel would be interested in their compound. They claimed that the compound was unstable and difficult to make. The Penfrel team wanted to see for themselves and they thought they could use their knowledge of the chemistry to improve on the compound. Gorman started to waffle. They said that maybe they would develop the compound themselves. Daniel immediately suspected some internal strife for the naysayer there. After working with them for over 6 months, Daniel and his team gave up and started their own program starting with Gorman's compound and trying to improve it themselves. Gorman never progressed the compound or the series further and they never outlicensed the program. Even though Daniel's team made good progress with the project, they were never able to get the compound all the way through clinical trials because of toxic reactions in early studies.

This experience is typical. The naysayers assume that everyone would agree with their reasoning in killing the program including outsiders. Of course, that is not always the case and when such a prospect arises, the naysayer gets nervous.

On the other side of the naysayer is the scientist who cannot understand why his/her favorite project is being killed by the company. This inability or unwillingness to understand or agree with the corporate decision frequently leads to subterranean efforts by the scientist and their buddies to continue the project. Their hope is that they will discover something new that will change the corporate mind. Although Daniel saw both sides of this coin, in his experience, the naysayers came out on top and the scientists ended up like Quixote until they reached a dead end or finally realized the futility of their pursuit. Of course, sometimes it is these very scientists who are the program champions that we so desperately need to move programs forward in spite of the naysayers.

For program champions, on the other hand, the risk rises exponentially as the compound advances later and later into development. Why? Because the farther the compound advances, the more expensive the project becomes. A phase three trial for an antibiotic usually runs around thirty million dollars. Program champions (sometimes the champion is more than one person) have to continue to point out every reason why the compound or program should go forward. They have to speak about the medical need, the market prospects, how the trial strategy is well informed, etc, etc. They have to continually keep the positive aspects of the program in front of the eyes of top management. If the program champion is "wrong," and there is a late stage failure – they are the most visible image of that failure in the company.

But without these brave souls, the program champions, Daniel was convinced that few or no drugs would ever be developed in large pharmaceutical companies because there is little or no risk in saying no.

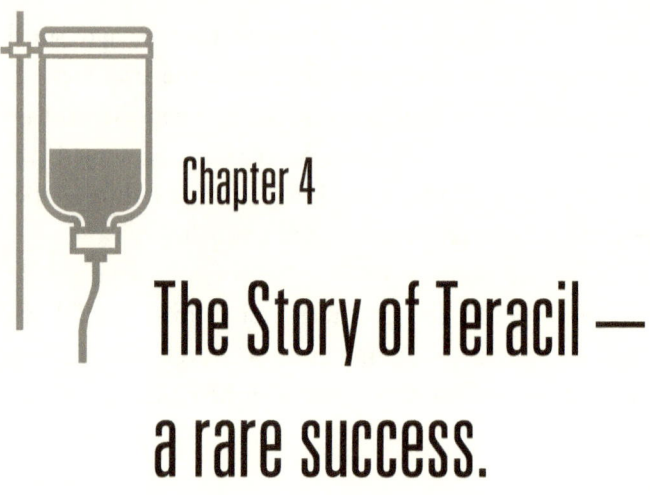

Chapter 4

The Story of Teracil — a rare success.

After working on the antiviral program, Daniel then tackled the antibiotic discovery program. Penfrel was marketing peracillin, a compound that Daniel had studied as an academician. He even helped direct some of the clinical trials that led to its approval early on. When Daniel arrived at Penfrel as the VP for infectious disease research, Penfrel had a struggling antibiotic discovery program where the first molecule to emerge had failed in early clinical trials. The team was now tasked with coming up with a backup molecule to try again. Of course, one could question why they had not chosen a backup earlier in the program – but they didn't. They also had a second antibiotic program that they wanted to move from the laboratory into clinical development. Neither was going to be easy.

The story of the failed antibiotic and the struggle to identify a suitable backup is the story of antibiotics in the pharmaceutical industry today. The failed drug was called TMC-TMG, perhaps appropriately so since the chemical name of the compound was so long and so hard to pronounce. TMC-TMG came out of a fascinating program put into place in the early 1990s by Tom Franklin when he first arrived at Lederman Pharma. Tom loved the tetracycline antibiotics. He felt that they were broad spectrum – that is – they were active against a

30

wide array of bacterial pathogens – and that they had been in clinical use for 45 years so that clinicians understood how to use them and were comfortable treating patients with them. The problem with the tetracyclines was that, with 45 years of use in animals, on crops, and for human infections, most bacterial pathogens were resistant. Lederman Pharma, on the other hand, discovered tetracyclines in the late 1940s and still marketed one, girocycline. Lederman's scientists had extensive experience and deep expertise in tetracycline chemistry and had a large library of tetracycline derivatives to study. Tom tasked his chemists with identifying new tetracycline derivatives that would be active against the resistant strains.

To understand a little about the challenge posed by Tom, it is important to understand something about antibiotics in general and about tetracycline resistance specifically. Antibiotics are toxins or poisons – just not to humans or human cells. They specifically kill bacteria. But, like humans, bacteria are living things and their genes and proteins, while very distantly related to human genes and proteins, are not 100% unrelated. So finding molecules that kill bacteria specifically by interfering with key bacterial processes but that don't effect human cells is already a pretty big challenge.

Tetracyclines inhibit the synthesis of proteins in bacteria but not in human or animal cells. They do this by binding at a crucial site on the protein assembly machine called the ribosome. Luckily, bacterial ribosomes are different from human ribosomes. This explains why antibiotics might inhibit bacterial protein synthesis but not that of human cells. Bacteria become resistant to tetracyclines in two major ways. Most commonly, they find a way to pump the antibiotic out of the cell. In this way, the concentration inside never gets high enough to allow much antibiotic to bind to the ribosome. These pumps are encoded by so called jumping genes – pieces of DNA that can be transferred from one bacterium to another with ease. The other major mechanism of resistance is a protein that actually binds the ribosome and prevents the antibiotic from binding – a so-called ribosomal protection mechanism. These proteins are also encoded by jumping genes. The fact that these resistance mechanisms can be so easily transmitted from one bacterium to another accounts for the widespread resistance to tetracyclines.

So, Tom was asking his chemists to come up with a new molecule that would avoid both the efflux pumps and the ribosomal protection proteins carried by resistant bacteria. Since, on a molecular basis, these resistance mechanisms were completely unrelated to each other, this was a pretty tall order.

To start with, the microbiologists at Lederman Pharma constructed a series of different bacterial strains carrying, either singly or in combination, most of the known genes encoding tetracycline resistance. In this way, new molecules coming from the chemists could be assayed for their ability to inhibit the growth of the resistant strains. For molecules that worked, the microbiologists also made sure that they were not toxic to human cells and that they, in fact, inhibited protein synthesis carried out by bacterial ribosomes but not that performed by human ribosomes. Eventually, they would show exactly how the new molecules bound to the ribosomes using highly sophisticated techniques like x-ray crystallography where you can visualize individual atoms.

Very early on, the chemists found that by modifying one position on the tetracycline molecule, they could begin to overcome resistance by both resistance mechanisms. It turns out that the way this worked was simply by increasing the strength of binding of the ribosome by the tetracycline. In this way, the tetracycline would bind the ribosome before the efflux pump could suck it out and the ribosomal protection protein could not prevent binding of the new tetracyclines. After only a few years of work, they had a large series of molecules that worked very well. They called the new series the teracils. Then they had to pick the best one for clinical development. That is where things can start to go wrong.

Because testing for toxicity in animals is expensive, Lederman Pharma decided to pick the best compound based on activity against bacteria and its cost of manufacturing. None of the compounds in the series had been tested for toxicity in animals as yet. It turns out that tetracyclines are hard to make. Actually, tetracyclines are natural products. That is, they are manufactured by a bacterium called *Streptomyces aureofacium*. The organism actually manufactures a number of tetracycline molecules, some of which are good antibiotics and others are not but are rather the building blocks for the final tetracycline antibiotics. But if you want to modify the tetracycline, you have to purify one of these molecules and subject it to various chemical reactions in the laboratory. At one point Daniel had to try and identify the cheapest source in the world for the tetracycline that was the starting point for synthesis of the backup molecule his team finally chose. Guess where it was? – at Penfrel in Montvale, New Jersey, right next to Daniel's office on the Penfrel campus. When it comes to making tetracyclines, no one knew more than Penfrel (that is, the Lederman Pharma group they acquired). The molecule originally chosen by the Lederman Pharma scientists, TMC-TMG, was a precursor to Lederman Pharma's marketed antibiotic, girocycline. As such, it was less expensive than, say girocycline. It was primarily for this reason that TMC-TMG became the clinical candidate rather than any of the other active compounds in the series.

TMC-TMG then had to be manufactured to a high quality and at large scale to allow for the toxicity testing in animals and for additional testing in animal models of infection. One of the great things about antibiotics is that the animal models of infection are so reliably predictive of how the antibiotic will work in humans. From studying how an antibiotic works against infections in mice, scientists can frequently make a very good guess as to the dose needed to treat humans. This is not the case for almost any other area of drug discovery. The experiments undertaken in mice at this point then provided a good estimate for the dose would be needed to actually treat human infections. The toxicity testing in animals showed how much antibiotic would be toxic in animals and provided information on potential toxicity for humans. The dose difference between the efficacious dose (the dose required to treat infections in animals successfully) and the toxic dose is known as the safety margin. This is important, because when you first start dosing a new drug in humans, you have to know where (at what dose) to start. These animal studies give you a guide. One usually starts at some small fraction of the dose where there is no toxic effect in animals. The dose in human volunteers is gradually increased until some toxicity or intolerance is identified. Hopefully, that point will be a dose that is greater than the projected efficacious dose in people. In the case of TMC-TMG, that was not the case. The Lederman Pharma scientists projected that they would need a dose of about 300 mg per day to treat infections in people. But the drug was not tolerated at that dose. The highest dose that could be tolerated by the volunteers was less than half that. The problem was that, even though the drug was given intravenously, it caused nausea and vomiting as the dose increased beyond 100 mg per day. Even with anti-nausea medication, the volunteers could not tolerate higher doses. So, that was the end of TMC-TMG.

In fact, tetracyclines are known to cause gastrointestinal upset including some nausea and occasional vomiting at doses used in human therapy. But the frequency of these side effects is low for the tetracyclines on the market. Even though TMC-TMG was an intravenous drug targeting seriously ill, hospitalized patients, the intolerance would have precluded any sort of wide usage. This experience would poison the well of the Penfrel management for other compounds that might come out of the new tetracycline series.

Daniel arrived at Penfrel about a year after the death of TMC-TMG. The scientists at Penfrel had finally started to look at the animal toxicity of other compounds in the series, compared, of course, to TMC-TMG and tetracyclines in clinical use. In order to carry out these experiments, they had to convince the Head of drug safety and metabolism (DSM) at Penfrel that the tetracycline program was not dead with the demise of TMC-TMG and that additional animal testing and its attendant expense was worth it. That conversation occurred between one of the key Penfrel microbiologists, Richard Noland, and the senior VP for drug safety, Ralph George, one dark and rainy night in a trailer.

Richard was of medium height, stocky, with a head of curly dark hair. He always wore a kind of ironic half-smile as if he were in another world looking in at you through some window. Daniel would walk one floor down from his office to the laboratories where the scientists had their offices and stop in to chat with them on a regular basis. Whenever he would stop by to speak with Richard, Richard would barely look up from his computer, would continue to tap away, and yet would have an intelligent and responsive conversation with Daniel. For Daniel, at first, this was very unsettling. With time, Daniel realized that Richard was a master of multi-tasking. He had a photographic memory and could absorb and respond to multiple inputs at the same time. Richard was a brilliant scientist who had trained with a Nobel Prize runner-up at Rockefeller University. He wrote a cookbook (cheesecake was a particular specialty) and had written screenplays for a science fiction series that were actually broadcast. Richard was also the lead scientist on the tetracycline project and it fell to him to speak with Ralph George.

Ralph was tall, had a graying beard, rather thin and balding with a long face and distinct nose that was almost caricature-like. He had a quiet voice, but spoke with confidence and determination. There was no one at Penfrel who could kill a project more quickly than Ralph. He was reputed to be a connoisseur of fine wines – something he indulged in to excess according to many. He ran his department in autocratic fashion such that all decisions, even relatively minor ones, had his imprimatur. This was frequently problematic since Penfrel had many projects going on simultaneously and Ralph could not keep up with the progress of all of them – so things were left aside waiting for his decision on many occasions. Daniel learned from Bob Stein that to get things done, he would have to have frequent conversations with Ralph just to keep the infectious diseases projects high on Ralph's priority list. Of course, the other department heads tried to do the same – but somehow, Ralph and Daniel established a rapport that helped to keep Daniel's projects going.

Penfrel was building a new animal facility in upstate New York not far from the Adirondacks National Park, a five-hour drive from Montvale. Daniel was sure that one reason for Penfrel's choice of location for the facility was to make it harder for the animal activists to get near the place. You could get there by car, or by small planes going to Vermont or by taking the New York-Montreal shuttle. Daniel preferred the latter staying in a hotel overnight in Montreal and renting a car the next day. Richard always drove up.

One stormy night in April, Richard took to the road heading for upstate New York. When Richard arrived the rain was pelting down and there were frequent lightning strikes on the mountains nearby. At the site, the executives and everyone else had temporary offices in trailers. It looked like a mobile home park with a big hole in the ground in the middle of it. Richard ran through the rain in the dark searching for Ralph's trailer office. When he finally arrived, he was soaked. They started their conversation with Richard dripping all over Ralph's floor. Ralph insisted that the TMC-TMG program was dead and that Lazarus would not rise. Richard argued that the series of active compounds was a large one and that several very active compounds had not been chosen originally because of their higher projected cost of goods. Ralph said, "See – that's what I mean. Even if you find one that is less toxic, its cost of goods will be prohibitive."

"We don't know that. For all we know, we will be able to charge a higher price or the cost of goods could come down substantially once we get to a commercial scale."

Ralph finally agreed to a small, carefully constructed toxicology program to study several potential backups to TMC-TMG. But Richard was sure that no matter what the outcome of the study, Ralph would never be a supporter of a new tetracycline clinical program. He would be proven wrong.

When Daniel arrived at Penfrel, the data from this small study were in. And they weren't very helpful. Tetracyclines typically cause certain problems when given at high dose to animals. One problem is that they cause the release of histamines. This is like causing a severe asthma attack. This does not occur in humans, at least at the normal doses. So this toxic effect is pretty irrelevant to humans. The second major effect in animals is on the bone marrow where tetracyclines tend to cause anemia and decreased production of other blood cells. Again, this effect does not really occur in humans either. So, when comparing the toxicity of different tetracyclines, you end up looking at effects that would be important if you were going to treat rats for their infections, but not if you planned to treat people. The problem in people, nausea and vomiting, does not occur in animals. There is no good model for this side effect in animals, at least not for the tetracyclines.

Nevertheless, there were several molecules in the collection of 14 studied that appeared to have a better toxicity profile in rats than the others. Daniel and his team then pulled out all the data on activity in test tubes and in animal models of infection on all 14. They quickly narrowed the list down to two of the new tetracycline molecules that had activity even better than TMC-TMG and seemed less toxic in the rat. Their next step was to study these molecules in a more definitive animal model of infection so they could better define their activity against infection and so they could estimate what the efficacious dose might be in humans. The results of these studies unleashed a current of excitement in the group. They knew that, for the two backups, the dose in humans might be as low as 100 mg per day – three times lower than what was projected for TMC-TMG. What they didn't know was which one of the two to choose. To make their choice, they were again forced to have a conversation with Ralph. The team would have to ask Ralph and his group to carry out formal toxicology studies on both compounds in order to choose one to take into human volunteers. As it turned out, when Ralph saw the new data showing how well the compounds worked on infections in animals, he readily agreed.

Ralph requested that the team try and understand the mechanism for the nausea and vomiting caused by tetracyclines in general and the new compounds in particular. This turned out to be an impossible task. There were two animal models for vomiting that had been used for cancer drugs. One was in ferrets and the other in pigeons. Giving a drug intravenously to pigeons, which is the way you had to deliver the new derivatives from infectious disease research, was virtually impossible. The ferrets could not tolerate even small doses of the drugs, but they did not vomit or have other gastrointestinal side effects as far as the veterinarians could determine. Penfrel's scientists never could determine the mechanism for the nausea caused by the tetracyclines.

Of course, even after the agreement by Ralph and the drug safety group, Daniel's team still had to get the agreement of the rest of the executive committee at Penfrel. Testing two compounds in animals meant spending money, and anytime you wanted to spend money at Penfrel, you had to have the approval of some high level committee.

When Daniel first arrived at Penfrel, it had a system of spending requests whereby even people at the vice president level (like Daniel) had to get the CEO to sign off on expenditures of $25,000 or more. Understandable (not) since Penfrel was only bringing in something like $10 billion at the time. Of course, before it got to the CEO's desk, it had to be signed by Daniel's boss, his boss, the site head at Montvale, the Chief Financial Officer and who knows whom else. All Daniel knew was that it took months. Even the executive committees that had to approve moving molecules forward in the development process were more efficient than capital and contractual expenditures at Penfrel. Penfrel consultants constantly complained about the length of time it took Penfrel to pay them.

To move a compound forward at Penfrel, the team of biologists and chemists worked with the help of a project manager and representatives of all the different functions involved in the project. The list of team members was long and included the chemists and biologists involved in discovering the molecule, the chemical manufacturing group, the drug product folks (who take the chemical powder [called API] and make it into a pill or into an intravenous product, or into a pediatric syrup, etc.), the drug safety people, the pharmacologists who study how the drug is absorbed and excreted in animals and people and finally the clinicians who would eventually be charged with studying the drug in people in clinical trials. A marketing person was usually assigned to these teams early on to assure that, in fact, there will be a market for the product eventually and that the company was not wasting its time. In fact, considerable effort was spent making sure that the science behind the new product jived with both the clinical needs of patients and physicians and also with the commercial needs of the company. These two things hopefully go together – but for example – for drugs for the developing world like those for TB and malaria, there is a huge disconnect between the medical need and the commercial opportunity - hence the existence of the Gates Foundation and other non-profits.

The job of the project manager is to make sure that the team sets a reasonable time frame in which to accomplish all its tasks, that key team members such as the manufacturing and drug safety folks, work to meet the timelines set by the team, and to represent the team on a day to day basis to various company executives. This is a tough job and these people are highly under-valued by many in the industry. Those on the discovery side of things do not want their baby represented by someone with no scientific standing in their area of expertise, a common failing among project managers. The other problem was that many decisions were made in the day-to-day conversations between the project managers and company executives outside the team. This infuriated everyone else on the team. On the other hand, the industry does not function well without someone to unite the team in a disciplined way around timelines for what is always a complicated project with many moving parts.

Infectious disease projects were particularly challenging since the scientists must stay deeply involved with the product all through development and even after launching the drug onto the marketplace. This is because an antibiotic requires constant support by the microbiologists to ascertain how the activity of the antibiotic holds as the antibiotic is used. Regulatory authorities require that companies monitor pathogens for resistance to their marketed drugs fearing that wide use will lead to increasing resistance. The microbiologists also have to assure that hospitals and clinics will be able to test the pathogens they isolate from patient samples for the activity of any new antibiotic being tested in clinical trials or being launched onto the marketplace. This requires the development and manufacture of test kits. The microbiologists at the pharmaceutical company usually lead this process.

To actually begin to get more resources to support the further development of a compound at Penfrel, the team had to jump through a number of bureaucratic hurdles. The first required convincing the other preclinical managers, that is Daniel's peers that headed the discovery efforts for cancer, neurosciences, cardiovascular, arthritis, immunology and the head of chemistry that the data supporting the compound was scientifically strong and that the compound was, in their opinion, worth further investment. With that endorsement, the team then had to present to the executive committee. This group included the head of Research and Development (Alan Smith's boss), the heads of various functional areas like chemical manufacturing, Ralph George (head of drug safety) and the heads of the US and European commercial groups. Only with the endorsement of this last committee would the company support carrying out the large-scale manufacturing and the toxicology studies that would be required to begin to test the compound in humans.

At the executive committee called to review the infectious diseases team's proposal to carry out two formal toxicology studies on two potential TMC-TMG backups, many of the Penfrel executives were shocked. The head of manufacturing was particularly incensed since his group would have to manufacture large amounts of two difficult compounds that he had no confidence would be successful. No one had proposed anything like this previously. But, luckily, this was the way the industry was going at the time. The pharmaceutical industry was starting to realize that the chances of failure early in development were high and that they needed to have identified backup compounds early and to have tested them sufficiently so that they could quickly replace the lead compound in development if needed. Alan Smith, was a big supporter of this approach and he won the day for infectious disease. Of course, another month or two was lost preparing for all these meetings – including the meeting with Alan to get his support prior to the executive committee meeting.

After the executive committee meeting, the team then had to get the infectious diseases chemistry team together with the manufacturing chemistry team. Before you can do anything to study a compound, somebody has to make enough of it at high enough quality. This transition is a fascinating one. It turns out that the chemical reactions used by the chemist at the bench early on, in order to make say 100 milligrams of compound, are frequently completely different from the reactions used by the manufacturing chemist to make kilograms of compound at high quality. And frequently, these two groups of chemists do not get along well, do not interact well and don't work together well. Daniel found himself between the head of the infectious diseases chemists and the VP for manufacturing on more than one occasion just to get them to play nicely in the sandbox together. He hadn't signed up for this! Daniel was shocked and surprised that these problems would arise since everyone at Penfrel, theoretically, wanted projects to succeed.

The other thing the chemists had to do was to be sure that all their patents were in place so that both these molecules and all the other ones the chemists had made and many they didn't make were covered. If this wasn't in place, someone could come out with another molecule using similar chemistry and take what market Penfrel might have garnered. In fact, that's what happened a few years after Penfrel's molecule went into late stage clinical development. Someone slipped! A tetracycline derivative that differed from Penfrel's ultimate lead compound by only a few atoms was pushed into clinical development at a biotech company. This particular structure was somehow not covered by any of the many Penfrel patents. Of course Penfrel itself had lots of projects where other company's molecules were targeted by trying to get around their patents. This practice is nothing new in the industry, but you are always shocked when it happens to one of your babies.

After about six months, Daniel's team finally got the results on the formal toxicity testing of the two backups. One was clearly more toxic to animals than the other, while both were about equally active against infections both in the test tube and in animal models of infection. They obviously chose the less toxic backup to bring into clinical trials. It was called PEN-639 later to become teracil. 639 represented the last three digits of the Penfrel chemical library number for the compound.

About this time, Penfrel underwent a major reorganization. There would be three of them during Daniel's six years there. The significance for Teracil was that the drug chosen to go into clinical trials by Daniel's infectious diseases group and the Penfrel project team was required to once again get permission of the executive committee to move the compound forward. But the entire process and the executive committee changed with this latest reorganization.

As part of the reorganization, all compounds going into clinical trials had to be prioritized based on their potential commercial value. Daniel was asked to collaborate with the marketing department at Penfrel to determine the potential value of a drug that had never been tested in humans. He wanted to laugh when he heard this news since it seemed like a joke. But his management, no longer Alan Smith, was quite serious. So, Daniel and his team met with marketing to draw up a plan to ascertain the value of PEN-639.

First of all, since the overall chance of a drug making it to market starting from the point of departure for PEN-639 was 20%, any potential value would be corrected for this risk – that is - reduced 5-fold. Then, of course, the marketing people wanted lots of information that Daniel and his team could not provide with any certainty. How often would it be dosed? Once a day is good. Twice a day is OK. Three times a day not so good, etc. Well, Daniel pointed out, based on studies in rats and monkeys, their best guess was once per day. The marketing people wrote down once per day. Are there any side effects?

"The rats get asthma symptoms and anemia at high doses but this won't occur in humans."

"What about nausea or vomiting?"

"We don't see it in animals."

The marketing team wrote down – no side effects.

"What about spectrum of activity. What bacteria does it cover in humans?"

"Well, in the test tube it is active against all kinds of bacteria except maybe two or three human pathogens."

The marketing team wrote down – broad spectrum. They like broad spectrum.

"Is it active against resistant bacteria?"

"Are you kidding? That's how we made it, by screening for activity against resistant bacteria!"

Daniel was starting to get a good feeling about this.

A few weeks later the marketing researcher brought back the results of their studies. She said, "There are lots of broad-spectrum antibiotics out there. Most are generic and are sold for pennies."

"But almost none are active against the resistant strains. This will be an intravenous antibiotic sold in hospitals. You can charge a high price." Daniel replied.

The marketing team went back to their research. Weeks later they came back to the PEN-639 project team with a positive signal to move forward. They said that the drug value was bigger than a breadbox but was not a Cadillac. Daniel looked heavenward as if asking for help.

Daniel, being a scientist, could not understand how such important decisions within the company could be based on the least reliable sort of data. The market estimates for a drug that has never been studied in people are nothing better than semi-educated guesses. There is little science involved and certainly very little reliable data. It was a situation rife for manipulation either by the scientific team who wanted the project to move forward or by management who might, for a variety of reasons, not want it to move forward. Plus, since the prioritization process was just that, it was competitive. The various therapeutic areas were competing for scarce Penfrel dollars to support the favorite projects of each therapeutic area. Thus, the different areas could criticize the projects of the other areas. Of course, this was a game of limited return since it invited retaliation.

PEN-639 still had to win approval of the executive committee that would compare the market estimate for PEN-639 with that for other drugs in the Penfrel pipeline. At the committee meeting there was much sniping. Daniel was furious when the new head of clinical development for infectious diseases, who had not said anything negative about PEN prior to the meeting, suddenly piped up with – "We should go forward cautiously since the original lead compound caused nausea and vomiting at subtherapeutic doses." Of course, he might have been right. Daniel knew it. Daniel knew that you could not predict this side effect from the animal models and he knew that PEN-639 might fall by the wayside during its early clinical trials. The only difference that provided hope was that the predicted therapeutic dose for PEN was three times lower than that for its predecessor. The head of manufacturing also sniped complaining that the starting material for manufacture was expensive and that there were six difficult and expensive steps of synthesis from the starting material. But, to everyone's surprise, Ralph George, the head of drug safety, was supportive. "The animal data do not predict any lack of safety in humans and PEN-639 is much more active than the original lead. If we want a backup, this is a good one." PEN-639 was blessed by the executive committee and would move cautiously ahead into its first trials in human volunteers.

The clinical team was preparing for the first trials of PEN-639. The first question they ran across was how to blind the study. When you are studying a compound in human volunteers, the trial is run so that nobody knows who is getting what. At each dose level, usually 6-8 patients receive the drug and 2 receive a placebo (salt water or sugar water). In this way, the side effects of the drug can be separated from those of the procedure or the solution the drug is dissolved in for example. In the case of PEN-639, the patients would receive an intravenous injection of drug or a salt solution similar to the salt concentration in human blood. The first problem facing the clinicians was that PEN, being a tetracycline, was golden in color. So it would be obvious to all in the special clinical unit where the study would be performed which patient was getting saline and which PEN. The team decided to cover the IV bags with a brown paper bag and to use green colored tubing from the IV bag to the patient's arm to mask the color of the PEN. This was not a perfect solution, but seemed acceptable to everyone including the FDA.

The next problem occurred in the third or fourth dose group. When carrying out this trial, one usually starts with single doses of drug. So for example, a group of 6 patients might get an infusion of 10 mg of drug with two receiving saline. If the drug was well tolerated and no lab abnormalities indicated anemia or liver or kidney damage, the clinicians would go to the next group where subjects might receive 25 mg of drug. By the time the team got to 100 mg, several patients were already complaining about nausea and there were a few episodes of vomiting. Even with just single doses, by 2-300 mg, everyone was nauseated or vomiting. Surprisingly, so were a couple of the patients who just got saline.

"Why?", Daniel asked the clinical pharmacologist in charge of the study.

The clinicians reply was enlightening. "Well, all the patients are in a single room, separated by curtains that are not always closed. They can see each other and hear their discussions with clinic personnel. They can hear other subjects vomiting. Nausea is contagious under these circumstances. How would you feel if your neighbor was vomiting?"

All the patients were getting the drug in a fasted state. The Penfrel clinicans also looked at the effect of food on drug tolerance and on absorption of the drug. In patients fed a large breakfast, there was no interference with absorption of the intravenous drug (no surprise) but there was a significant amelioration of the nausea and vomiting associated with giving the drug. The nausea was also helped by giving the drug in two daily doses instead of once per day. The team also learned that older subjects tolerated the drug much better than the young volunteers used in the first studies.

Because Penfrel management was so concerned about this side effect, yet another executive committee was convened to review the results of the single dose studies before they would authorize proceeding to a multiple dose study. This delayed things again for a period of several months while data was analyzed and reports were prepared. A multiple dose study in volunteers was approved and carried out and resulted in data showing that doses above 100 mg per day were poorly tolerated, but lower doses seemed OK. Of interest, the Penfrel clinicians were much more concerned about the side effects of nausea and vomiting than the clinicians actually administering the drug to the volunteers. The clinicians in the study center understood that the side effect was not dangerous. The problem for the executive committee was that no health volunteer had been able to complete ten days of dosing at the dose needed for adequate therapy, 100 mg per day.

Finally, after much debate, the executive committee approved going into a small trial of just 50 patients with skin infections to see if the small doses of teracil that were better tolerated would actually work to cure the infection. They were encouraged by data showing that older subjects tolerated the drug better and that feeding also helped with the nausea. The team chose two doses, 25 mg twice per day and 50 mg twice per day to be studied in 25 patients each who had serious infections of the skin and subcutaneous tissues where staph and other Gram-positive pathogens were the most common culprits. This would be the first late stage clinical trial of a new antibiotic by Penfrel and the first in over 15 years by the old Lederman Pharma team. Daniel was especially concerned since he knew that 100 mg per day was probably the lower end of the dose that would be required to treat serious infections adequately. Everyone else, including the skeptics, was in a celebratory, optimistic mood. The study was completed within 6 months. Other than the two doses of teracil, there was no other standard antibiotic control. The study was run such that the clinicians at the hospitals treating the study patients did not know which dose the patients were receiving, but at Penfrel, Daniel and his team saw all the data in real time. By the end of the study it was clear that 100 mg per day divided into two doses would be effective in people with serious skin infections while 50 mg per day would not. On the other hand, about one third of those receiving the higher dose had at least some nausea, but none of the patients discontinued their treatment because of side effects. Daniel and his team knew that teracil would offer a new way of treatment for patients with serious infections caused by highly resistant bacterial pathogens.

To further verify this, Daniel knew that Penfrel would have to study teracil in more difficult infections caused by more difficult pathogens like the feared Gram-negatives. Physicians would not be ready to adopt teracil to treat these infections or even to test the drug in a large-scale trial in patients with these infections until there was some clinical data to suggest the drug would work there. Daniel and his team went back to the executive committee to once again request support for another small-scale trial to establish whether teracil would treat serious Gram-negative infections. Daniel and his team chose to study teracil in intra-abdominal infections where Gram-negative pathogens were dominant. These infections included infections of the gallbladder in patients with gallstones, patients with appendicitis where the appendix had already started to rupture, and other serious infections in the abdominal cavity. All of these patients would require surgery. Daniel was enthusiastic about these trials because he knew that at surgery good samples would be obtained for microbiological testing. The team would know which bacteria were causing the infections and which teracil could treat. This time the 100 mg per day dose of teracil would be compared to the best and most powerful antibiotic on the market for Gram-negative infections, imipenem. This meant that the trial would have to be blinded to everyone, including the clinicians and scientists at Penfrel. No one would know the results of the trial until several months after the trial had been completed. Teracil would have to look at least as good as imipenem and would have to be well tolerated by these very ill patients. So, another year would be required to start and complete this trial in serious intra-abdominal infection.

Three years after starting clinical trials of teracil in humans, Penfrel received the results of their study in patients with serious intra-abdominal infection. Teracil was just as good as imipenem and should work against even those strains resistant to imipenem. Not only did teracil work, but there was not any more nausea and vomiting among the teracil patients than among those who received imipenem. Daniel always thought that this was because patients with abdominal infections were likely to have nausea anyway and that this essentially masked this side effect of teracil.

At the same time, Penfrel was starting to get phone calls from physicians treating patients with infections where their pathogens were highly resistant to antibiotics available on the marketplace. They were requesting teracil on a compassionate basis. This question comes up during clinical trials for every new antibiotic that can potentially treat resistant bacteria for which no other treatment is available. Like most companies in this position, the Penfrel team had decided that they would not make their drug available for such requests as there are risks. Treatment of these patients, some of whom are very sick, could result in false signals of safety risks for the new drug just based on the desperate state of the patients being treated and unrelated to the antibiotic. But it wasn't the team who got these phone calls – it was Daniel. Daniel himself was on the other side of the coin as a physician back in Cleveland. At his hospital there was an outbreak of completely antibiotic resistant Gram-negative bacteria causing pneumonia in the surgical intensive care unit. At the time, imipenem was undergoing its late stage clinical trials. He requested imipenem from Merck, showed it was active against the Gram-negatives causing the outbreak, and was able to treat and save patients with these infections using imipenem on a compassionate basis. Daniel was unable to deny the physicians who called him requesting teracil. He overruled his own team. Teracil began to be used by clinicians around the world trying to treat patients with various infections caused by highly resistant pathogens.

One patient in particular remained with Daniel throughout the rest of his career. She was only in her 50s and had a congenital condition called spina bifida where her body never closed correctly around her spinal cord. These patients are subject to repeated infections that can invade the spinal cord, brain and the bloodstream. This particular woman had acquired an infection with a Gram-negative superbug called Acinetobacter that was resistant to every antibiotic available on the market except colistin. Colistin is an old antibiotic, discovered in the 1970s that was never studied very well, so no one knew how well it worked in serious infections. It routinely caused kidney damage. The patient had been treated with colistin, still had her infection and had suffered kidney damage. Her physician was requesting teracil. Penfrel sent teracil to the physician who made sure that the patient's Acinetobacter superbug was susceptible to the drug. It was. They then started intravenous therapy with teracil. Within two weeks the infection had cleared. Because the bone of the vertebral bodies had become infected, the physician prescribed a six-week course of treatment with teracil. At the end of the six weeks there was no further evidence of infection, and a year later, the patient was doing well. For Daniel, this would be his personal poster child for why he went to work in the pharmaceutical industry.

At this point, the teracil team had drawn up plans and a budget taking teracil through initial phase III trials, to the FDA and the European authorities for approval and finally to market. The budget for these trials was going to be well over $100 million. Additional studies would be carried out after approval to expand the market. Once again the team would have to return to the executive committee to present these plans and obtain approval to go to this very expensive step. At Penfrel, there was a so-called blockbuster mentality. Penfrel would not invest in phase III trials unless the peak year sales of a drug were projected to be at least $500 million to one billion dollars. Of course, this is rare for an injectable antibiotic. But, in fact, the antibiotic that had already been developed and marketed by Lederman Pharma would attain this level of sales before its patent expired. Daniel and the marketing team worked furiously on developing a commercial model for teracil sales.

When Daniel first moved to Penfrel, he immediately thought of his old friend, Roger Butler. Daniel and Roger had known each other for over 20 years. Roger was tall, stocky and rapidly losing his light colored hair. Their families were close, having met in Cleveland when they were students. They were at each other's weddings. Roger worked for Lederman Pharma in the sales group before moving to Glaxo to do market research. Daniel was able to lure Roger to Penfrel's market research group shortly after Daniel's arrival at Penfrel. They quickly formed a strong working relationship and sometimes worked separately from the rest of the marketing team. This became problematic for Roger's management and ultimately would make his life at Penfrel impossible. When Daniel had to work with the commercial group to estimate teracil's market potential, he wanted to enlist Roger on the project. But the head of infectious disease marketing blocked this move. It turns out that people are people and petty jealousies, envies and ambitions stand in the way in the pharmaceutical industry just like everywhere else. The Penfrel marketing group developed a forecast suggesting that teracil's peak year sales would reach the one billion dollar mark. These projections are usually based on the number of patients you think you will be able to treat and the price you will be able to charge in different countries. Of course, these data are all based on projections based on interviews with small numbers of experts – who are not those who prescribe most of the therapies you are talking about anyway. Infectious disease physicians, for example, who are frequently interviewed in such surveys, don't prescribe nearly as much antibiotic as say internists or general practitioners. Roger and Daniel set out to establish their own, secret, independent model. They came up with a figure more like $500 million assuming some good luck. Theirs was not the model presented to the executive committee. But with the commercial group's support, the data on teracil in skin infections and in intraabdominal infections, and given the projected expenses to complete the data package for the regulatory authorities, the executive committee agreed to allow the drug to proceed into phase III trials.

Next, the team would have to meet with the regulatory authorities in the US (the FDA) and in Europe (the EMA). These authorities would have to agree on the design of the large clinical trials that Penfrel planned to carry out and submit for market approval after completion. The FDA was worse than Penfrel's executive committee.

Clinical trials of antibiotics are not like those for most other medicines. For most other drugs, you need to show that the drug is superior to no therapy (or at least no drug therapy). But since we know that antibiotics work to treat infections and are safe, it is not ethical to withhold safe and effective therapy from patients with a serious disease. Therefore, you can only study antibiotics compared to an already marketed antibiotic that has been previously shown to work. Since the days of the sulfonamides in the 1930s, no antibiotic has ever been directly compared to no therapy. So all the antibiotics marketed today, including the generic antibiotics, were shown to work when compared to another antibiotic that had been compared to yet an earlier antibiotic, etc. Statisticians hate this. What you have to do is to show that your new antibiotic is "not inferior" to a marketed antibiotic. Since, statistically, you can never show that one drug is exactly the same as another, and it is not possible to study enough patients to show that your drug is better than the 90% success of the older drug, you have to show that it is not inferior – within some statistical margin.

One problem is that clinical trials are not real life. To enroll a patient in a trial you have to inform the patient and or their family of the risks of the trial and the new drug. You then have certain inclusion and exclusion criteria. You might not want to enroll patients that have certain other medical problems that might increase the safety risk for the patient. You can't enroll patients so sick that they might not survive long enough to evaluate whether the new drug works or is safe. In real life, there are no inclusion and exclusion criteria. Patients need treatment and as a physician you do the best you can.

But, if there are resistant infections occurring, why can't you just show that your drug works there and let it go at that? At the time of teracil, a number of other companies tried to study their antibiotics by including resistant infections in the context of their non-inferiority trials. To obtain a label from the FDA that stated that the drug was active against the resistant infection, you only needed to enroll 20 or so such patients. But, in fact, the inclusion and exclusion criteria normally used successfully excluded most of those patients and many companies failed to get the resistance label for their new antibiotic. In fact, the regulatory authorities are now finding ways to allow companies to study antibiotics in highly resistant infections – but at the time of teracil, such pathways did not exist. A non-inferiority study was the only choice. But it is the statistical margin for these studies that is the problem.

The statistical margin around non-inferiority defines how close the new drug has to perform to the old drug. The closer you want the two drugs, the more patients that would be required. Antibiotic trials, until teracil, usually used margins of fifteen per-cent. The Penfrel team suggested a set of seven trials that would involve the standard statistical margin of fifteen per-cent and would require about 2700 patients with either skin or intra-abdominal infection or pneumonia. The FDA proposed similar studies at a ten per-cent margin requiring over 6000 patients. This difference required more than twice the patients. This meant the trials would be twice as costly and take longer to complete. That in turn meant that the drug would be delayed in reaching the physicians and patients that needed it. It also meant that Penfrel would be deprived of sales for an extra year or so therefore pushing back Penfrel's ability to make a return on its research investment in teracil. When the results of the FDA interaction were again presented to the Penfrel executive committee, Penfrel management simply refused to develop the drug under these circumstances. Daniel and his team were completely discouraged. Teracil was on its deathbed.

Daniel started calling colleagues in academia and in other pharmaceutical companies in search of ideas on what to do. There is a trade organization in the US called PhRMA or the Pharmaceutical Research and Manufacturing Association. It had a committee called the antimicrobial working group that dealt with problems such as this – those issues that would be common to the entire industry. Daniel was able to join this group and presented the problem to them. At the time, most large pharmaceutical companies had active programs to discover and develop antibiotics. He was told that a number of companies had been told the same thing by the FDA. They would have to increase the statistical stringency of their trials – more patients, more money, more time. The infectious diseases departments at other companies were all coming under pressure from their management since the money manager types were afraid that the increased costs required by the FDA would mean that the companies might not ever see a return on their investment. They all agreed to help. While these companies competed with each other, they also had common cause in dealing with the increased stringency now being demanded by the FDA. There was a great deal of enthusiasm for doing something – the question was – what to do.

At the same time, Daniel contacted the Infectious Diseases Society of America. This group represented infectious diseases physicians throughout the US and included many who were involved in clinical trials of antibiotics. As an infectious diseases physician, Daniel had been an active member of the society. These physicians needed new antibiotics to deal with the ever-increasing numbers of resistant infections that they were facing. They also agreed to help – but asked the same question – what do we do? Daniel, the IDSA and PhRMA decided that the best next step would be to have a workshop (sponsored by PhRMA and IDSA) with the FDA to explore their reasoning behind insisting on these more stringent and costly trial designs and to see if there was some way to address their concerns in a less costly manner.

The FDA agreed to the proposed workshop. Daniel was appointed to present the case for industry. The workshop was scheduled in several months. But before the workshop, PhRMA and FDA scheduled a pre-meeting to try and understand the FDA's position. At this meeting, the head of the statistics group at FDA walked in and said simply that the pharmaceutical industry, in their trials of antibiotics over the years, had been "getting away with murder." What he meant was that he felt that the non-inferiority trials that had to be undertaken were not statistically reliable indicators that the antibiotics actually worked. He wanted to see comparisons to placebo. Daniel and other infectious diseases physicians in the room were so shocked they could barely speak. How could you enroll a patient with a serious infection in a trial of a new antibiotic when they might, in fact, be randomized to get no treatment? It was unimaginable that someone would actually propose such a thing.

The meeting continued after this initial outburst, but the PhRMA team remained shocked. The FDA explained the basis of their concern, though. It had to do with something the FDA called biocreep. If you are always comparing one drug to another, and, potentially, one drug could be as much as fifteen per-cent worse than the other (the margin), eventually, you might get to a point where the latest drug was no better than no drug at all. This phenomenon became known as biocreep. Of course, this argument turned out to be complete nonsense and Daniel was sure that the FDA statisticians knew it was nonsense.

Daniel and a team from PhRMA met weekly to discuss their presentation and to put slides together. There were also several meetings with IDSA to discuss progress. Biocreep became a key focus of the discussions. The infectious diseases physicians who use antibiotics all the time knew that the FDA had never approved an antibiotic that did not work against infections caused by bacteria that were killed by the drug in a test tube. There were other problems over the years related to safety or to the fact that a new antibiotic might not provide enough of an advantage over older, cheaper drugs such that it did not sell well. But the physicians were convinced that antibiotics cured infection and were better than no drug at all. The statisticians in PhRMA agreed. Their reasoning was that, in fact, the FDA required two trials for a drug to be approved. The statisticians could show that at the normal fifteen per-cent margins used previously by FDA, if you did two independent trials showing that the new drug was within that margin compared to the old drug, the chance of approving a new drug that was even ten per-cent worse than the old drug was less than three per-cent. So much for biocreep.

Daniel was a little lost in the statistical details, but he understood the bottom line. The FDA requirements represented an unreasonable and unlikely concern that threatened the very existence of antibiotic research and development within pharmaceutical companies around the world. He assumed that the statisticians at the FDA had gone mad and that, on the battlefield of FDA management, they had won a decisive victory. Daniel was already hearing rumblings of discontent from other pharmaceutical companies. In fact, within a few weeks,

Roche announced that they would no longer pursue antibiotic research and development citing the fact that it would no longer provide a sufficient return on investment for them. With that announcement, Daniel's worst fears were confirmed. He knew that the workshop and negotiation process would be too slow and would probably not be enough to contain the FDA statisticians. Daniel started calling the press.

A year prior to the FDA onslaught, Businessweek carried a story on the discovery of new antibiotics in the pharmaceutical industry. Daniel and teracil were featured and Daniel had established a good relationship with the journalist who wrote the story. He reconnected with her with a description of the latest moves by FDA. He also called contacts at the Financial Times, Wall Street Journal, New York Times and others. Of course, Penfrel had a strict policy forbidding any press contact by employees in the absence of the Penfrel public relations department. Daniel carried out his campaign in secret, making all the phone calls from home and hoping that the reporters would succeed in keeping his name out of their stories. Most wrote stories highlighting the problems caused by the FDA's stance and quoting the FDA as saying that the increased stringency was required to assure that the drugs were safe and effective. Daniel's name was never mentioned. But Daniel did author an article published in a prestigious infectious diseases journal suggesting that the FDA's stance would kill the already impoverished antibiotics effort within the pharmaceutical industry.

The day prior to the scheduled workshop, the teams from PhRMA and IDSA met in Rockville, Maryland, near the FDA offices, to rehearse their presentations and make any final adjustments that might be required. The IDSA physicians focused on presenting the medical need for new antibiotics that would be active against resistant pathogens. They emphasized the paradoxical situation for antibiotics. Because physicians were afraid that using a new antibiotic would select for resistance, they reserved it for use only when the infection was likely to require the new antibiotic – in other words, when resistance to older antibiotics was likely. But for the industry, this meant lower sales and a slower uptake in sales for their new drug – anathema to pharmaceutical companies and their shareholders. The IDSA then discussed the idea that trials done under the older designs (fifteen per-cent margin, smaller patient numbers) were valid as far as they were concerned. The PhRMA team demonstrated the increased costs and time of trials that the FDA was now requiring, showed that the statistics for the older designs was adequate and would prevent so-called biocreep, and made a plea for a balanced approach that would allow companies to make a return on their investment in antibiotics and still provide safe and effective therapy for the patients who needed it because of resistant infections. The FDA statisticians presented their counter-arguments suggesting that there was no way to know whether modern antibiotics were any better than no therapy at all.

As luck would have it, the director of the anti-infectives department at FDA, Moses (Moe) Weinstein, was an experienced, extremely intelligent and politically savvy FDA staffer. He saw the conundrum facing the industry as well as patients and physicians. He also had no love for the statisticians in his own department. He presented a variety of approaches open to companies wanting to develop antibiotics, and, in his conclusion, he stated that there was never an FDA policy requiring that antibiotic trials follow the more stringent design that many companies were facing at FDA. Daniel, the PhRMA team and the IDSA clinicians were cross-eyed. Every company team coming to the FDA for the last year had been told that they would now have to follow the more stringent design. But at that point, there was nothing to do but accept Moe's word and start over with the FDA.

Daniel and the Penfrel team did just that. They arranged another meeting with the FDA reproposing their original design and waited for feedback. This came in the form of a faxed response from FDA just prior to their scheduled meeting. In the fax, the FDA restated their position that the more stringent trials would be required. Daniel quickly called Moe. Moe himself came to the meeting with Penfrel that occurred the next week. He overruled his own team and allowed Penfrel to go ahead with the less stringent design they had proposed originally. The delay for teracil caused by this process with the FDA was another year.

At the same time, Lilly and Bristol-Myers-Squibb announced that they too would abandon antibiotic research and development. It turned out that Penfrel would be the last company to be allowed to use the less stringent design for antibiotic trials. Everyone else would have to use more patients and spend more money.

Daniel immediately called the new president of research and development at Penfrel with the news. His name was Sam Stern – known not so fondly as Yosemite Sam for his habit of "shooting from the hip." On this day, Daniel was apparently lucky since the shot was aimed at the FDA and not at Daniel. Sam immediately agreed to allow the teracil trials to go forward without convening yet another executive committee meeting. Sam said he would let his managers know of his decision himself.

Teracil and TMC-TMG were both discovered at Penfrel in 1992 and the patent was approved in 1994. Nine years had passed between its discovery and its go ahead for its final trials that would not even start for another year. It would not reach the market before 2005 – 13 years after its discovery. In fact, this length of time is not beyond the industry average, but is certainly longer than normal for antibiotics prior to the year 2000 when the FDA began its descent into neverland.

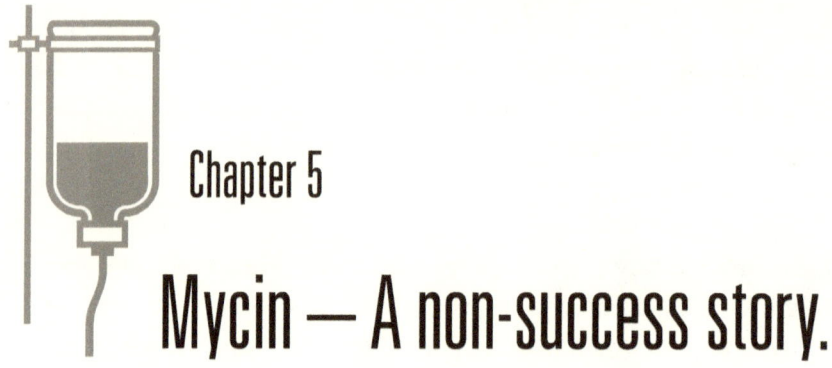

Chapter 5

Mycin — A non-success story.

After Daniel had completed his reviews of ongoing projects at Penfrel, he asked his scientists to come up with ideas for future directions. One direction Daniel wanted to go was back into the past. Bacteria exist in nature in a competitive natural environment. Nutrients are limiting. They must compete to survive. Over the eons of evolution, bacteria acquired means of attacking competitors – antibiotics – and means of defending themselves against these poisons – resistance. Most of the antibiotics marketed during the 1940s, 50s and 60s were such natural products either completely natural or optimized through additional chemistry. These included the tetracyclines, erythromycin, penicillins and cephalosporins. To find these, companies went around the world collecting soil samples and other materials where they thought bacteria might be competing within a natural environment. They then isolated these bacteria, cultured them on various media, and looked for substances that would inhibit the growth of other bacteria. They would then try and isolate and characterize these molecules – antibiotics. Many of the large pharmaceutical companies had large collections of such natural products in various stages of characterization. And many of these had been discarded because they were toxic to human cells or because they did not fit whatever strategy the company was pursuing at the time. For example, in the early days, most companies wanted antibiotics that killed all sorts of bacteria – so called broad-spectrum drugs. But in our modern world, the worst and most widespread resistant bacteria was a Gram-positive superbug called MRSA – a highly resistant form of staph.

A drug active against Gram-positive bacteria would be acceptable to a company today (maybe) when it was not 20 years before.

Daniel knew that Penfrel, through its acquisition of Lederman Pharma, had one of the oldest and largest natural product collections in the world. He asked the infectious disease scientists to go back through the entire collection looking for compounds that were discarded but which might be candidates for development in today's changed world. They quickly put together a list of possible profiles that would allow them to go through all the old data and they set about reviewing the entire collection of tens of thousands of extracts and compounds. After several months of work, they came up with only a few possibles. One was called BA-45. No one knew the significance of the acronym, but BA-45 was remarkable. The extract was derived by growing a bacterium isolated from soil in media, then filtering out the bacteria such that a sterile extract was left. This extract killed other bacteria – but only Gram-positive bacteria like staph and the superbugs, MRSA and VRE (an almost totally resistant bacteria called, enterococcus). Daniel couldn't believe that they had had this kind of luck. And remarkably – the extract, when injected into infected mice, would cure the infection and allow the mice to survive. BA-45 was a rare jewel indeed.

But BA-45 was, in fact, a mixture of compounds. The scientists did not know the structures of the compounds in the mixture nor did they know which might be active and which not. Luckily, the technology for determining structure in the 1990s was far superior to what it was in the 1960s when BA-45 was discovered. Within another month or two, the natural product chemists at Penfrel had deciphered the structures of the components of BA-45, purified them and showed that all were active with some being more active than others. The chemical differences between those that were more active already gave the chemists a direction in order to further optimize BA-45 through chemistry to derive a pure and even more active compound.

Daniel and his team were crawling the ceiling with excitement over this project. There was so much to do. How did BA-45 kill bacteria?

How did it work in mice? Was it going to be toxic to animals or people? What is the best way to carry out the chemistry to get greater activity but avoid toxicity? To answer these questions and carry out all these tasks would take several years and a fairly large team including a number of chemists. In order to garner resources like chemists that were outside the infectious disease team, Daniel would have to convince Alan Smith and his colleagues that BA-45 was worth the effort. At that time, Alan and the rest of Daniel's colleagues could make their decision without involving marketing and the clinical development group. Daniel was able to convince everyone there that the MRSA threat was enough to justify putting some effort into BA-45 and he gave his team just one year to come up with a candidate for development. This fast a timeline was without precedent in Penfrel and probably, along with the argument for medical need, won the day for Daniel and his team. Since Daniel was a physician, his word seemed to carry extra weight with Alan and his scientist colleagues at Penfrel. This certainly helped smooth the way for the infectious disease team at least early on.

In fact, the chemists rapidly found a way to purify the key components of BA-45, separate out a key chemical starting material and to alter it such that it could be modified to improve activity. The chemical doors were open. The next problem was the producing bacteria. The problem was that to get several hundred milligrams of the BA-45 starting material, the natural products microbiologists would have to grow at least a hundred liters of bacteria. This was not going to be practical since the chemists would need grams of starting material at a time. Daniel met with the natural products microbiologists to discuss what could be done. They were very accustomed to this problem and pointed out that it was going to be a simple matter of strain selection. They would subject the bacterial strain to mutagenesis with DNA damaging compounds. These compounds would cause mutations in the bacteria and some of these mutations would increase the production of BA-45. They would repeat the process until they had a strain that could produce enough compound for the chemists. They also proposed another approach where they would identify the genes in the producing bacteria responsible for synthesizing BA-45. These genes would be isolated and used to make a strain that produced BA-45 at high levels. In this case, both approaches worked quickly and the natural products microbiologists could produce grams of BA-45 from a hundred liters of culture. The chemistry program was off and running.

The compounds were rapidly tested for toxicity against mammalian cells, for activity against bacteria focusing on MRSA and for activity in the mouse model of infection. A smaller group within Daniel's department worked on trying to understand how BA-45 killed bacteria. Yet a different group worked on how the compounds worked in mice. This led to their first roadblock.

Both for the antibacterial research and for the antiviral research in the Penfrel infectious disease team, the scientists were frequently delayed having to wait for tests of mouse blood needed to understand how compounds were either working or not in animal models of infection. The drug safety group, under the supervision of Ralph George, carried out these tests. The drug safety group had no particular incentive to work on the infectious disease projects and samples frequently sat in that department for weeks. Daniel spoke to Ralph on several occasions and Ralph always promised to speed things along. This strategy worked for one set of samples at a time.

Daniel had had enough. He gathered his team of biochemists and asked them to set up their own facility to test blood from mice and other animals so that they could avoid the constant delays in the drug safety group. They found money to buy the equipment they needed, had the scientists trained and hired two new scientists with expertise in animal work. Within six months they had their own testing systems established and working. Daniel was in fact concerned that Ralph George and his team would resent the apparently duplicate effort within infectious disease – but the opposite was true. The drug safety scientists helped the infectious disease team set up their assays and systems and were ultimately happy to accept and use the data coming from the infectious disease scientists. The BA-45 project sped along as did a number of antiviral projects that had also been languishing in the basement of drug safety.

Within less than a year, the infectious disease team had identified a series of derivatives of BA-45 with much improved activity against MRSA both in the test tube and in mice. They understood how to avoid toxicity to human cells while improving antibacterial activity and they understood how the drug worked. It inhibited a key reaction in the synthesis of the bacterial cell wall and they even understood exactly how the reaction was inhibited. BA-45 bound to an intermediate in the pathway and prevented the synthetic pathway from proceeding beyond that step. Human cells do not make a cell wall – so the activity was specific for bacteria. They also understood how BA-45 worked in animals. It depended on the total dose given during a period of time. This means that the drug could be given once a day, twice a day or many times in a day and the activity in mice (and in people) would not be affected. Since some toxic effects might be due to very high concentrations, increasing the number of doses could reduce toxicity without reducing the antibacterial effect of the drug in treatment.

The team decided that it was time to test the compounds for toxicity in animals – but they wanted to get a quick look at this themselves before they asked drug safety to get involved. They used the most active compound they had that did not cause toxicity in human cells and simply looked first at single doses in mice and then later at a ten-day treatment course in mice. They found that the drug caused kidney failure in the mice after about 5–7 days of treatment. But this did not stop anyone. The team went back to the chemical drawing board and found another derivative with similar anti-bacterial activity that was also not toxic to human cells. They carried out the same test in mice and this time the test was clean at even higher doses than they had used for the first compound and at doses several times those needed to cure infections in mice. They thought that this compound would be the one to take into human trials.

Now it would be time to present to the Penfrel executive committee. This meant first getting the marketing group involved to estimate potential sales for such a compound. It also meant convincing Ralph George that the compound was worth testing in rats and dogs for toxicity such that it could be presented to the FDA to get clinical trials started in human volunteers. Daniel and his team went into this presentation with all the confidence in the world. BA-45 got a new name – Mycin. But the road for mycin would be anything but well paved.

Ib Hassad was the head of the manufacturing group for Penfrel. Ib was of Syrian descent, but had been living in the west for most of his life. He was a chemist trained in England before moving to the pharmaceutical industry more than twenty-five years previously. Even before the presentation got beyond showing the chemical structure of Mycin, Ib was on his feet. The compound is now purified by chromatography, he said. We can't manufacture kilograms much less tons of mycin using chromatography. We have to find a completely new synthetic method. Not only that, but the producer strain is still not efficient enough for large-scale manufacturing. We will have to continue to work on improving production of the starting material from the producing strain. All of this means that the cost of goods right now is going to be extraordinarily high and we do not know if we will be able to improve it before we get into larger tox studies and clinical trials. If we can't bring down the cost of goods, mycin will never be profitable.

Then it was the turn of the commercial group. Mark Mann, the head of market research got up. He said, Let's look at the emerging competitive picture. For therapies for staph infection including MRSA, we have linezolid from Pharmacia in late stage trials, we have oritvancin, dalbavancin and daptomycin in late stage trials, we have teicoplanin being sold in most countries outside the US and we have vancomycin on the market as a generic and inexpensive drug. I don't see a way to differentiate mycin from any of these. How will we compete?

Daniel and his team did not have good answers to any of these questions. Mycin was just coming out of the laboratory and they were already being asked for both a manufacturing and a marketing plan. While this seemed unreasonable at the time, Daniel learned that he had better start thinking about these things very early in the drug discovery process – perhaps even before starting out on a research path. But to Daniel's surprise, the overall head of research and development at the time was Alan Smith's boss, George Finkel. George had interviewed Daniel before he was offered the job at Penfrel. George was an outstanding scientist in his own right having been one of the discoverers of how cholesterol was synthesized in the body. He won the Lasker Prize in medicine for his work before he joined industry. Penfrel was his third job in industry. George was also a man with enormous patience and a large perspective – he saw the big picture. "First," he said, "its much too early to worry about large scale manufacture of the product. Lets see if we can get beyond early toxicology studies – you can work on strain improvement and purification of the compound as we go along, Ib. As far as the market goes, most of the compounds you discuss are still in clinical trials. We don't know if they will make it to the market or not. Even though it is likely that a couple will make it, its too early to know whether we will have a competitive advantage or not. Lets go ahead to the next step and think about competitive advantage over the time it takes to get to the next step – that should be almost a year from now – no?"

Daniel was dumbfounded by the wisdom of George's approach and by the fact that he was so easily able to overrule his team without seeming to make anyone angry or resentful. They all kind of shrugged their shoulders and accepted his ruling deciding to just get on with it and see what would happen. That is, all but Mark Mann from marketing, who would stick to his guns for the duration. And, of course, George himself would be the victim of reorganization arriving without warning.

As it turned out, only linezolid and daptomycin would make it to the market within the decade following the discovery of mycin. The rest would spend the next ten to fifteen years foundering in trials and struggling with an FDA that changed its goal posts at every opportunity. George's long-term view was the right one.

In the meantime, everyone had their marching orders. Daniel's team would focus on the microbiology to show that mycin would be active against strains that were resistant to the pipeline drugs underlined by marketing. Daniel's team was already aware of the new, experimental antibiotics in clinical trials, but they just had not focused on them from a competitive point of view as yet. They would also work on more definitive animal models of infection to allow them to predict the dose that would work in people and to define the minimum effective dose in mice. The latter would help establish a safety window once the toxicology studies were completed. The ratio of the dose where there was no adverse effect in the animals to the minimum effective dose would be their safety window.

Ib Hassad and the manufacturing group would start working with the chemists on the mycin team to improve the producing strain and get a rapid, if expensive way forward to manufacture enough drug to carry out all the necessary studies. Ralph George and the drug safety group would design, and when enough drug of high enough purity was available, would carry out the initial toxicology studies required. All these activities would take the better part of the next nine to twelve months. Mycin was off and running – but where?

The first sign of trouble came from Ralph George in drug safety. Mycin had to be administered intravenously to the rats used in the early toxicological testing. For this the veterinarians usually used the tail veins of the rats. But they were having trouble because after the first few doses, the tail veins were clotting and could no longer be used. Daniel, Richard Noland and others on the team wondered if there could be a problem with the solution used to deliver the drug. This was just mycin dissolved in saline. Could it be that the drug was not dissolved correctly? Was there a problem with the acidity of the solution that was irritating to the vein? Neither of these explanations were correct so they had to assume it was something about the drug that was causing the problem. Could this ultimately be corrected by dissolving the drug in something that would protect the veins of humans? Since human veins were so much larger anyway, and therefore the drug would be diluted much more quickly in humans, everyone felt that this problem with mouse tail veins should not stop the progress of mycin. So they used a catheter to cannulate a larger vein in the rats to deliver the drug. This allowed the toxicology studies to progress.

In these studies, different groups of rats would be given either saline (salt water) alone or saline with mycin at three different dose levels for about a week. Since this would not be the definitive toxicological study leading directly to human trials, an abbreviated protocol was used. At the beginning and end of the study, blood would be obtained to look at blood counts and blood chemistry and the animals would be sacrificed and their organs examined. If anything would be visible grossly by the naked eye, the organs would also be examined microscopically.

The next call from Ralph was not good news. The animals showed clotting in arteries and veins, including the large vein used to deliver the drug. There were small clots in the lungs and even in the hearts and brains of the animals. It was amazing that the animals continued to look healthy during the week of study and that their blood counts and chemistries showed nothing amiss. But the clots occurred even at the lowest dose used – and this dose was lower than the dose that could successfully cure animals of a staph infection.

Just as Daniel, Richard and the team were digesting Ralph's news, Mark Mann called to say that they had been working on a market projection for mycin. He sent it over. It suggested peak year sales of no more than $100 million. To arrive at this figure his team assumed that all the drugs in late stage trials would succeed and would make it to market before mycin. It also assumed that mycin would have little to no advantage over the competition. Mark Mann's group said that if the drug could be given in pill form, their analysis would have been much more favorable. But that their research with infectious disease physicians indicated that another intravenous drug targeting staph infections was not seen as an important medical need in spite of the growing number of such infections and the growing use of vancomycin, the only drug on the market at the time that could be used for serious MRSA infections. One good thing was that Mark Mann had supplied a mathematical model in the form of a spreadsheet that could be manipulated by Daniel's team. They tried various scenarios where only a more limited number of drugs got to the marketplace and where mycin had an advantage over staph resistant to those drugs. But the fact was that Richard Noland's own research had already shown that there could be cross-resistance between several of the new drugs including mycin and that this could emerge, based on the frequencies that they saw in the test tube, in patients on therapy.

Richard Noland argued that they might be able to overcome the toxic effect of the drug by delivering it more slowly or by diluting the drug or both. He argued that it was likely to be a local effect of the drug in high concentration on the rat veins. He also stated that it was not likely to occur in human veins which were larger and where the drug would be diluted more quickly following injection. But to win the day, they would have to overcome the data on resistance that they had generated within their own group, convince the veterinarians to continue to study a drug that they thought was toxic and to convince the marketing group that their research was incorrect and premature. Not only that, but Daniel found that he was now having trouble convincing himself that mycin was going to be an important addition for patients and their physicians. He began to think that, in this case, mycin would have to die.

When a drug is approved for toxicological testing by the Penfrel executive committee and fails, it must be re-presented by the team that championed the drug and that team must itself make the proposal to discontinue development. In this case, Daniel, Richard Noland and the rest of the scientists would have to swallow a bitter pill. The decision to discontinue would mean that all experimental work except what might be needed to finish up a few things for publication would have to be stopped. They did – sort of. The team made its proposal to discontinue development and this recommendation was accepted by the executive committee without discussion. But Richard Noland and his group continued work on mycin trying to keep their efforts secret from Daniel. They tried the approach of slower administration and diluting the drug solution before administering it to see if they could avoid the problem of systemic clots in rats. They even tried some other derivatives of BA-45. Of course, Daniel knew exactly what was going on because these secrets do not keep. Daniel saw all the supply budgets. Ultimately, mycin was the subject of a fair number of very interesting scientific publications, but never became a drug. Mycin followed the fate of 95% of everything else discovered in the industry – it was a scientific success but never even got into human trials.

As for the predictions of Mark Mann, several of the drugs that his group assumed would be competitors with Mycin either never made it to the market at all or would only arrive ten years after his model assumed they would arrive. They would not have been competitors and Mycin might have enjoyed market success if it could have ever been developed. But so it goes with the science and the non-science of drug development.

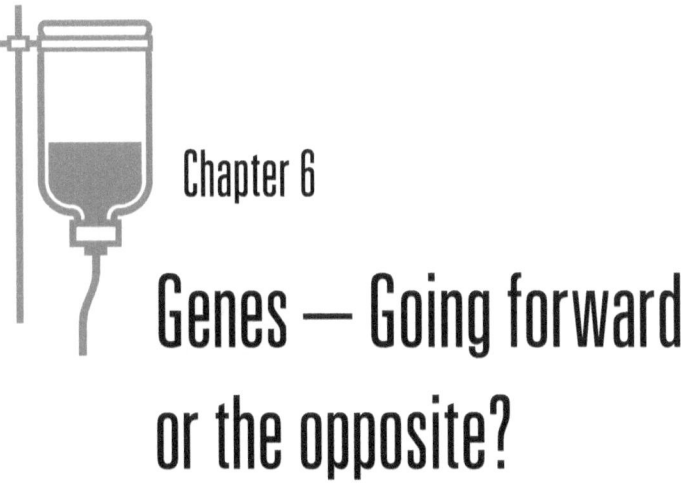

Chapter 6

Genes — Going forward or the opposite?

When Daniel arrived at Penfrel, he was aware that Smith-Kline Beecham had already started dissecting bacterial genomes looking for new ways to find antibiotics. He also knew about Bacto in Boston and their approach to the function of bacterial genes since he had just turned down their job offer so he could join Penfrel. Bacto had hired George Oldman in the meantime. George was a well-known and highly respected bacterial geneticist from Georgia. George had rapidly taken control of the research on bacterial genes at Bacto and was leading the charge to find pharmaceutical company partners for Bacto's research program directed at using bacterial genes to find new antibiotics. Daniel invited George to bring the Bacto team to Montvale, New Jersey to discuss Bacto's approach with Richard Noland and the Penfrel scientists.

The basic idea was to find genes in bacteria where, without the gene present, the bacteria would be unable to grow even in media containing lots of nutrients. Then, these so-called essential genes would be compared, based on their DNA sequence, to human genes or genes of other organisms to be sure that the genes would be unique to bacteria. Finally, the sequence of these essential genes would be compared among different bacteria to see whether they were unique to just the species being used in the research or would be found in many or all species of bacteria. The idea was that if you found a gene that was not shared with humans but that was shared with all other bacteria, and you could find a drug that could inhibit the function of that gene in bacteria, you could find a wonder-antibiotic that might not be toxic to people.

Looking back on this process, Daniel was constantly amazed at how everyone in the industry, himself included, could have been so naïve. But at the time, this was an exciting new approach to antibiotic discovery. And George Oldman was not only a clever geneticist, but he was also a good salesman. He was not tall, he was thin and came to meetings in jeans and sport shirts. The dress code for Penfrel executives required shirt and tie. In fact, at the Penfrel corporate headquarters in Madison, New Jersey, white shirts were preferred over colored shirts. But George would patiently walk up to the white board and draw out his explanations as to how he would carry out the genetic studies to dissect the bacterial genome, and then how the Bacto team would compare DNA sequences to those of other bacteria and to those in mammalian cells.

Daniel and Richard Noland spent many hours going over the Bacto proposals. They both agreed that the Bacto approach was going to be more revealing than any other available in the pharmaceutical industry at the time. Not only was that judgment wrong, but so was the entire theoretical basis for the pursuit of bacterial genes. If bacteria have been competing with each other over the eons of evolution, maybe all the antibiotics, or at least most antibiotics, that might work had already been discovered by the evolutionary process. It might be that the genes that might be uncovered within the genomes of bacteria were not accessible to inhibition by drugs. Not only that, but the scientists failed to appreciate the extent to which the bacteria could prevent entry of drugs into bacterial cells or the extent to which drugs could be pumped out such that the drugs, even if they might work against some bacterial enzyme, could never achieve a concentration inside the cell that would allow them to inhibit bacterial growth.

Many infectious disease research groups would purify the enzymes identified by their bacterial genetics programs as essential to bacterial growth. These enzymes would then be used to screen the large compound libraries of the various companies to find inhibitors. These inhibitors would then be optimized such that only infinitesimally small amounts of compound would be required to inhibit the enzyme. But if the compound could not enter and stay inside the bacterial cell to inhibit the enzyme, it would never work. And in fact, in virtually all of these cases, failure was the end result.

Another approach was to use bacterial cells that were engineered genetically to produce either a very tiny amount of the enzyme or a very great amount. These two bacterial strains could then also be used to screen libraries of compounds. In this case, the under-producing strain might be inhibited while the over-producing strain might not be- suggesting that the enzyme was the target of the compound. All this and more was done only to find out that none of the compounds could kill the bacteria because the compounds could not get in or stay in the bacterial cell itself or because the enzymes chosen based on bacterial genomics simply were not good targets for the drugs in the libraries of the pharmaceutical companies.

The pharmaceutical industry spent at least a decade banging their heads against this wall before realizing that we weren't yet ready for bacterial genomics. After five years, the Bacto-Penfrel collaboration would end with no compound ever having entered clinical trials. George Oldman was no less a scientist and no less a salesman after those five years. But he, like everyone else, was beaten by the bacterial cell membrane and the recalcitrance of the targets identified by the survey of bacterial genes.

During these years, Richard and Daniel never stopped searching for newly discovered bacterial genes they could target for new antibiotics. To do this, they voraciously read both literature and patents. They also kept in constant touch with their friends and colleagues in academia. One such researcher was Henrik Dortmund. Henrik worked with his wife Maria at the Rockefeller University.

To Daniel, the Rockefeller was the pinnacle of everything wonderful about academic research and at the same time, it displayed the worst of the dark underbelly of the academic world. The researchers there were held to a standard so high that only a few could actually survive there. They had to maintain at least two high-paying national grants – usually from the National Institutes of Health or the National Science Foundation. Only a few of the world's top scientists could consistently meet this standard. And the Rockefeller did not tolerate failure very well. If you lost even one of your grants, your time there to re-acquire your lost funding was limited to just a year or two. After that – sayonara. This policy led to fierce competition not only between researchers from the Rockefeller and the rest of the academic world, but also between researchers at the Rockefeller itself. The scientists there were constantly plotting to replace their competitors, take over their lab space or climb the administrative ladder over their colleagues' backs. Knives were always drawn if not always visible.

Daniel often wondered who could work under these conditions. But the scientists working there were the cream of American researchers – the top of the top. Henrik was no exception to this. His experiments were simple and at the same time elegant. His conclusions seemed obvious even when the concept being explored was highly complex. Richard knew that Henrik was working on a new bacterial gene that offered some promise of being a new target for antibiotics. He arranged a visit for himself and Daniel at Henrik's lab at the Rockefeller. By the time they arrived a month or so after their initial discussion, Henrik's paper on this gene had just been published in the Proceedings of the National Academy of Science. Daniel and Richard devoured the paper and their team analyzed the gene. In fact, this gene was critical to the synthesis of the bacterial cell wall. The most famous inhibitor of the synthesis of the bacterial cell wall was penicillin and all its related antibiotics. Since humans and animals don't have a cell wall, they don't have the genes or proteins necessary to make one. Inhibitors therefore tend to be effective against bacteria and non-toxic like penicillin. The bacterial cell wall remains the holy grail of antibiotic targets.

The analysis of Henrik's gene by the Penfrel team showed a major weakness – it was only present in one or two species of bacterial pathogens – certain streptococci. This meant that even if Penfrel were to identify a safe and effective inhibitor of the enzyme coded by this gene, it would only be active against these specific bacteria and no others. The physicians would have to know that the infecting pathogen was a particular type of streptococcus in order to use the new antibiotic. Daniel knew that this would never work since in at least 80% of cases physicians did not know what the infecting organism was when they started treatment and they frequently never knew. They needed antibiotics that could target an array of likely pathogens – not just one or two. But their appointment was on the books and Daniel was looking forward to an exciting day discussing science at the Rockefeller with Henrik and his team.

After a challenging drive from the Penfrel labs in Montvale, New Jersey to the Rockefeller on the East side of New York city, Richard and Daniel finally made it to Henrik's lab 30 minutes late around 10:30 am. Henrik greeted them in his office – a large, wood-paneled room lined with bookshelves filled with journals and volumes in a seemingly disordered array. There were two secretaries and a few young scientists working at small desks or sitting on several couches pouring over journals or notebooks. Henrik, Daniel, Richard and one of Henrik's post-doctoral scientists sat down at a small conference table with a laptop computer. Henrik himself presented the story of his discovery in a logical, step-by-step manner. Both Daniel and Richard were enthralled even though they knew much about Henrik's discovery already. They would ask questions from time to time. Richard, in particular, wanted to explore the remote possibility that Henrik's gene could actually exist in other species of bacteria. Daniel knew that Penfrel's gene experts were probably much better than Henrik's and that if this possibility existed, they would know. Henrik confirmed that he had no evidence that the gene existed outside of a couple of species of streptococci.

After two hours or so of discussing the science behind Henrik's gene, Henrik invited Daniel and Richard to lunch at a private table in the

Rockefeller dining room. Henrik spent a good deal of time over lunch exploring the extent of Penfrel's interest in his newly discovered gene. He had pointed out during his earlier presentation that he had filed patents to cover work on the new gene. Henrik, like most academics, thought that these genes could be patented. This concept was thrown out by the US Supreme Court decades later – but at the time, in spite of Daniel's skepticism, many scientists in academia and in industry were busy patenting bacterial genes. Henrik went on to laud the value of his discovery by noting that since an inhibitor of his gene would be specific only for one or two types of streptococci, it would be less likely to select for resistance in other bacteria. He suggested that he would be willing to license the rights to his gene to Penfrel for the sum of $1 million.

Daniel practically coughed up his chocolate cake. Richard's expression remained unchanged – not even a raised eyebrow. Daniel recovered quickly and explained his problem – that any antibiotic that they might discover that might actually work against Henrik's gene would have a spectrum of antibacterial activity that would be too narrow to be practical for physicians. That would mean that commercially, such a product would not be viable and therefore that Penfrel would have no interest in the gene. He neglected to point out that the figure mentioned by Henrik would have been too high in any case, even for an attractive bacterial target, by almost ten times.

Henrik's gene and his proposal became a legend at Penfrel, and years later, throughout the industry and even in academia. But this experience did not dissuade Daniel, Richard and the Penfrel team from continuing to try and find new ways to kill resistant bacteria. It also never dissuaded Henrik from trying to convince pharmaceutical companies to license his gene.

Another approach to find new antibiotics and to get around bacterial resistance was to look at those functions of bacteria that allowed them to attach and to invade the host – virulence functions. Some believed that this approach, by avoiding a drug that kills the bacteria directly, would not select for resistance. Daniel never believed this to be true since anytime bacterial survival is at stake, there will always be a few mutants around to escape. But virulence is still being pursued today in a number of companies.

The approach is not an easy one for several reasons. If all one does is inhibit the bacteria's ability to attach or invade, growth outside the body in media will not be inhibited. But in hospitals and clinics around the world, bacteria isolated from infected sites like abscesses, urine, blood and others are tested in the test tube to see which antibiotics work to kill the bacteria and which don't. Antibiotics are prescribed or prescriptions are corrected to take these results into account. But in the case of so-called virulence inhibitors, the bacteria won't be killed so there would be no way for a physician to know if his prescription was the right one or not. The doctors would have to accept on faith that the drug would work. This would require a complete change in the way doctors think about the modern treatment of infections.

Also, virulence factors are particularly specific to the species of bacteria involved. So there are virulence factors for staph, others for strep and others for certain Gram-negative bacteria. There are even specific virulence factors for specific bacteria important for invasion of only the human urinary tract. From the point of view of the practicing physician, this is problematic. When faced with an infection, doctors frequently don't know what the organism is that is causing the infection. How would something that is so specific be used in the real world?

Another problem with the virulence approach is that, even if you genetically completely obliterate a virulence function of bacteria, they can still cause infection – it just takes more bacteria to achieve this. Once an infection is already established, will a drug that inhibits attachment or invasion or some other virulence function be able to clear the infection by itself? Or will such a compound always have to be combined with an antibiotic? If the latter is true, how will it be studied in human infection to be sure that it will work in combination with an antibiotic that might already work pretty well?

In spite of these challenges, many companies were and are still attracted by this concept. Even Daniel, Richard and the Penfrel team made an attempt to find a virulence inhibitor. Richard was adamant that the approach could work if only they could choose the right enzyme to inhibit. Richard had in mind an enzyme that was required to produce many different attachment factors in many different bacteria hoping that any inhibitor that might be discovered would have a somewhat broad spectrum of activity. Daniel was finally convinced to try this in spite of all the obvious problems. But this project, although scientifically exciting and interesting, did not advance very far before Daniel left Penfrel and Penfrel pulled out of infectious disease research altogether. To date, no inhibitors of bacterial virulence have progressed to late stage clinical trials. The idea remains an interesting one, but, so far, not a practical one.

Chapter 7

BLI – One step forward . . .

As Daniel began to realize that the genomics effort with Bacto was not going to bear fruit, he began to worry that if teracil should fail, Penfrel would have no new antibiotics under his watch. He increased his efforts to find a way forward by looking outside Penfrel. Peracillin, the blockbuster antibiotic that Penfrel inherited from Lederman Pharma, was actually a combination of two drugs. The first was a penicillin-like drug, but the second was an inhibitor of bacterial resistance to the penicillins, the beta-lactamase enzymes. One defense used by bacteria in their natural environment is to develop enzymes that actually destroy an antibiotic. The enzymes that destroy penicillins and similar drugs are called B-lactamases. The first such enzyme was discovered by one of the key researchers working to bring penicillin into clinical use – Abraham Chain. And the discovery occurred years before penicillin was ever marketed. Gorman Ltd. was the first pharmaceutical company to develop a combination of a penicillin and an inhibitor of B-lactamases. This combination was the biggest selling antibiotic in history with over two billion dollars in peak year sales. Penfrel was not far behind with its own version of such a combination – but Penfrel's combination was directed at the more life-threatening and more resistant bacteria found in hospitals. Penfrel's drug targeted hospitals while Gorman's targeted infections occurring in the community. As such, there was plenty of room for both drugs on the marketplace and both were great successes. But the patent for both the Gorman antibiotic and for Penfrel's peracillin would run out in another few years. Penfrel wanted a replacement. So far, teracil was their best hope. But could Daniel and his team find something else?

Daniel did just that. He was aware, because of his days consulting for Smith-Kline Beecham, that they had a very interesting program to identify new B-lactamase inhibitors. Some of this work had been published and as such was freely available, so Daniel did not have to break his confidentiality agreement with SKB to discuss this with his team. Daniel knew that the SKB scientists who went to work for Gorman brought their ideas with them and that Gorman had picked up on this program. Just like Penfrel, they were looking for something to follow their blockbuster drug. Daniel asked the head of infectious disease chemistry at Penfrel, Tawfik Madad, to undertake a patent search focusing on B-lactamase inhibitors and both SKB and Gorman.

Tawfik, like Ib Hassad from manufacturing, was of Syrian origin. He lived in Montreal for a number of years working for a small pharmaceutical company there. Alan Smith had recruited him to Penfrel just after Daniel had started at Penfrel. Tawfik was extremely talented, easy-going, had worked on antiviral programs in Montreal and was an ideal person to lead the infectious disease chemistry group at Montvale. He was stocky bordering on chubby with a balding head and red cheeks. He could have been Santa without the hair and beard.

Daniel walked over to Tawfik's office in the chemistry building next door. The chemistry building was built on an atrium model where the chemistry labs and offices were behind large glass windows surrounding a large, open atrium. Walking through the building was like going to a shopping mall. You were always attracted to colorful items in the windows and tempted to offer to buy something. When Daniel walked into his office, Tawfik pulled out a large file folder. He then pulled out a few pages from various patents showing what Gorman had been publishing over the last five years or so. They apparently had been working on a new B-lactamase inhibitor modeled on the old SKB program. But this inhibitor had a different structure and was active against more B-lactamases than anything seen previously. And the patents claimed that it worked to clear infection caused by highly resistant bacteria when combined with a penicillin like drug in animal models. But there was no news anywhere that these compounds had ever entered clinical trials. So, either Gorman was exceedingly good at keeping secrets or they had run into some problem with the compound.

Based on the chemical structures revealed by the Gorman patents, Tawfik and Daniel speculated that there might be a problem with drug stability. If the drug simply fell apart in storage or when exposed to plasma, it would be useless. With Tawfik at his side, Daniel picked up the phone and called Ben Dolan, the business development person at Penfrel assigned to infectious disease. He, with the rest of the business development group, was at the Penfrel site outside Philadelphia and was, in fact, just down the road from the Gorman facility. The businesss development group was responsible for Penfrel's deal-making activities. Ben worked for the oncology and infectious disease areas. Daniel arranged a meeting between Tawfik and himself and the business people to try and get a meeting set up with Gorman. Daniel and Tawfik's idea was to see if the Gorman compound was worth bringing in to Penfrel to see if it could be developed as a follow-on the peracillin. It would clearly be active against many peracillin-resistant bacteria and would, therefore, open up a larger market. It would, at the same time, provide an important benefit for physicians and patients struggling with these resistant infections.

The meeting with Gorman took place about a month later. The Gorman scientists, most of whom Daniel had known for many years already, presented their data on a series of compounds as described in the patents Daniel and Tawfik had reviewed. They pointed out the problems they were having with drug stability as the Penfrel scientists had suspected. They said that because of this, Gorman had stopped work on the project. Daniel suggested that perhaps it would be worth letting the Penfrel scientists have a look. The meeting then took a turn towards the business terms that might be put in place to allow the Penfrel scientists to begin studying the Gorman compounds. After some preliminary discussion, the meeting broke up with a promise that the business people from both companies would follow-up soon and hammer out an agreement.

When nothing happened six weeks later, Daniel called Ben to find out what was going on. Radio silence is going on, Ben responded. Daniel called his contact at Gorman. Reading between many lines, Daniel thought he understood the problem. If the Gorman scientists allowed Penfrel to look at their drug and Penfrel found a way to solve the stability problem so it could be developed, the Gorman scientists would look incompetent. This agreement would never happen.

Daniel called Tawfik. "OK," he said. If Gorman won't play ball, what are our options?"

"Well, we could take their compound, make it, and use it as lead to modify the structure. We would have to do this in a way to avoid their patents."

"Can you do it? Is it legal to make their patented compound?"

"Of course its legal. You can do almost anything you want for research. If you try to commercialize something covered by someone else's patent – that's another story."

"We'll need an entire chemical strategy with an early test to see if we can make this work."

"I'll get back to you in a week or so."

A week later, Daniel was in Tawfik's office with Richard Noland and some of Tawfik's chemists. They presented the Gorman structure and explained the areas around the structure covered by their patents. They showed where the Penfrel scientists would have room to maneuver to modify the structure. Not only that, but Tawfik felt that the modifications he was proposing would make the compound more stable and would still maintain the better activity of the original Gorman compound. But there was a problem. Tawfik and his chemists had never worked on these structures before – they were called penems and they looked a little like the penicillins themselves. The chemistry was already difficult, and Tawfik worried that it would take considerable time for his team to figure out how to carry out the reactions necessary to make the molecules he was proposing.

"But without a chemical strategy, we can't move forward with this." Daniel complained.

"I know."

But then lightening struck Daniel. Twice a year, whether he wanted to or not, Daniel had to visit the Penfrel affiliate outside of Tokyo, Japan to review their work on teracil. Daniel hated these trips because the Japanese scientist and managers would always seem to agree to everything while Daniel was there and then do nothing during the next six months. It reminded Daniel of the movie, Groundhog Day with Bill Murray. Daniel just reasoned that the Japanese couldn't say no, especially to someone they thought was their hierarchical superior, so they nodded their heads to indicate that they understood what was being said but not necessarily to agree to anything. But one good thing about these trips was that Daniel got to know the chemists in Penfrel Japan. They were working on their own program trying to find penem (related to penicillin) antibiotics. They knew penem chemistry like the back of their hands. In addition, Penfrel was in the throes of deciding to shut down the entire Penfrel Japan facility and was looking for a buyer for the site. The chemists were about to lose their jobs anyway.

Daniel turned to Tawfik. "Lets get the Penfrel Japan chemists on this project. We'll find a way to pick up their salaries for the next year or so while you work with them and your guys can learn the chemistry to drive the program forward."

Tawfik agreed to champion this cause within his management in the chemistry group at Penfrel. He was successful, and another month later they launched the project with the help of the chemists at Penfrel Japan. They were magicians. They knew just which reactions to use when and where. New structures started to arrive in Daniel's labs for testing – and they were all active against resistant bacteria when combined with a penicillin like drug. All of them. Not only that, but they were making rapid progress on the problem of stability as well. Plus, the chemists at Montvale were starting to have success using the methods they were learning from their Japanese colleagues. Daniel and Tawfik agreed that they would promise Yosemite Sam and the Penfrel management a new B-lactamase inhibitor as a candidate for clinical development as a follow-on to peracillin within one year. And they delivered.

But the compound would lie fallow at Penfrel for over five years since the delivery occurred a few months before Penfrel abandoned antibiotics research and development altogether. Penfrel was incapable of deciding whether to go forward with the new inhibitor or not for all those years. When Daniel would run into the few infectious diseases scientists left at Penfrel during those years, he would always ask, "Where is my B-lactamase inhibitor?" He always got the same response until almost seven years later. When Penfrel finally did go into clinical trials, the one compound they tested failed because about a third of the volunteers suffered with a drug rash either during or shortly after completion of dosing. That was the end of the penem B-lactamase inhibitor program at Penfrel. Daniel was sure that if Penfrel had continued with their research on these inhibitors, they would have been able to find one that worked without toxicity and they would have been able to make it all the way to the market before anyone else. But as it was, with Penfrel's abandonment of antibiotics research, their B-lactamase inhibitor program would remain a failure. To Daniel, in a way, it didn't matter, since at that point, Daniel was deeply involved in another, even more promising B-lactamase inhibitor program that would make it all the way to market.

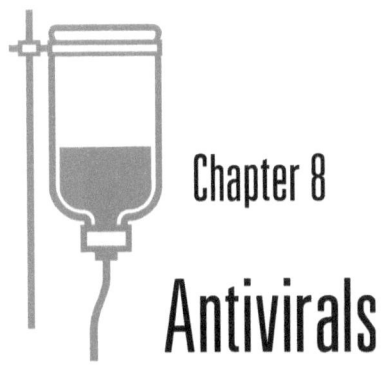

Chapter 8

Antivirals

Iain Cavendish was short, thin, had thinning reddish blond hair and spoke with a light, pleasant Scottish brogue. He was a good scientist trained in New York at New York University. He came to Penfrel to work under Sasha in Montvale New Jersey as a young man just after his post-doctoral training. Daniel was charmed. Also, Iain had been able to hold the team together after the Sasha's tragic and violent murder. After completing project reviews and getting the backing of Alan Smith and Penfrel's research management, Iain and the team dove into their remaining projects on Respiratory Syncytial Virus (RSV) and Cytomegalovirus (CMV). Daniel, at least early on, was very dependent on Iain since Daniel had never worked in the field of virology. Of course, Daniel was familiar with viral infections since he treated patients with AIDS, cytomegalovirus, influenza and other viral infections. But he had not thought about basic virology since his days as a PhD student in Cleveland. Luckily, he had Bos Killington available for consultation. Bos, early on, gave presentations discussing the medical need for RSV and CMV. He noted that a vaccine for RSV had been on the CDC's top priority list for new vaccines for decades. Bos said that the most pressing needs for an anti-RSV drug were infants who got severe respiratory infections caused by RSV every year during a well-defined RSV season. The other was immunocompromised patients – especially those with leukemia, those receiving bone-marrow transplants and those on chemotherapy who could suffer severe and frequently fatal infections caused by RSV.

Bos explained that virtually all humans get CMV infection during the first 5-10 years of life. But the virus remains dormant in the body and can cause trouble later if the immune system is compromised – especially by AIDS, cancer or chemotherapy. CMV infection was also a problem for newborns that acquired the infection from recently infected mothers who had not been infected as children, but became infected during late stages of pregnancy. The babies would get the virus while exiting the birth canal. The fetus could even get the infection in utero where it would cause widespread infection. According to Bos, these were the key medical needs for a CMV drug. Daniel realized that developing a drug to treat pregnant women or newborn infants would be challenging, but he wasn't yet ready to face those problems. First, they had to find a drug somehow.

Daniel learned that the way the Penfrel scientists searched for compounds that inhibited viral growth was most likely to identify those compounds that inhibited the very earliest stage of viral infection – the fusion of the viral membrane with the host cell membrane. This was a very specific process carried out by specially adapted proteins within the viral membrane. Penfrel's two lead compounds for RSV and CMV inhibited this process. The Penfrel group had also assured that the compounds they found were very specific for these viruses and their specific proteins in two ways. First, they showed that the compounds inhibited growth of only the virus of interest and not any others. This suggested that the compounds were not acting through some very subtle toxic effect on the host cell since that should result in the inhibition of other viruses as well. In the case of RSV, they were able to isolate a virus resistant to their lead compound and show that the mutations leading to resistance occurred in the fusion protein. This was both good and bad since it assured everyone that the compound was a very specific inhibitor but it also demonstrated that the virus could become resistant with therapy. RSV could also infect animals – but not easily. There were infection models in the mouse, the cotton rat and in two species of monkey where the cotton rat was the standard animal model. So the Penfrel scientists would have a way of testing their compounds in vivo in animals before going into human trials.

At the time of Daniel's arrival, the RSV compound was still being optimized by the chemists. But looking at the structure, Daniel knew right away that it would never be absorbed through the intestine – therefore it was most likely that it would have to be administered intravenously. Daniel felt that this would be acceptable since young children were hospitalized with life-threatening RSV infections every winter around the world. Adults also suffered colds and influenza like infections caused by RSV during the season and were occasionally hospitalized with bronchitis or pneumonia. But as the molecule was optimized and worked better and better in the test tube, the time for testing in animals arrived. The team elected to use the mouse model and the animal testing group within Daniel's department in Montvale got the model up and running. The first test was disappointing to say the least. In this model, the animals were infected with the virus and shortly thereafter were given the drug intravenously. The virus in the lungs would then be counted 24 hours later. If the drug were effective, the counts in the treated animals should be much lower than in the untreated control animals. But in this case, there was no difference between treated and untreated animals in the RSV counts. There was a difference in that treated animals suffered seizures and even death when the drug was administered. The RSV team had to go back to square one. But even looking at older, less active compounds with somewhat differing chemical structure, the lethal effect of the drug was still there.

At the time all of this was occurring, there was a drug used by physicians to treat desperately ill children and immunocompromised patients suffering from RSV infection. It was called ribavirin. It did not work well and had to be administered by inhalation of a nebulized spray in a tent. The drug crystallized on the skin and eyelids of the children and even on the skin of nurses and other caregivers in the room. Because it never worked well and posed a risk to healthcare workers (the drug could cause genetic mutations), its use was eventually discontinued by pediatricians. But during the time of the Penfrel RSV project, ribavirin was still used fairly frequently for very ill children. Knowing this, the Penfrel team decided to try administering their drug to mice by aerosol. To do this, they had to purchase special cages that would allow for delivery of such an aerosol but still protect the animal workers. They did this and the drug worked with no apparent harmful side effects to the mice. The Penfrel team then had the drug tested by academic colleagues in New York using the standard cotton rat model and aerosol delivery. It worked there as well.

Daniel was of two minds about this. He didn't think that aerosol delivery would be well accepted by pediatricians and other caregivers – but if this was their only alternative and the drug worked – why not? Daniel, Iain and the rest of the team began to examine how they could take this RSV drug into clinical trials in patients. First, they would have to establish that the drug was safe in doses that would be far above those that worked to cure infection. To do this, the drug would have to be tested in both a small animal, usually the rat, and in a large animal species. Because monkeys were smaller and would tolerate an aerosol delivery system better than dogs, and because systems for delivering aerosols to monkeys safely had already been developed within the biodefense sector, the monkey would have to be the large animal used. To do all this, the team would have to determine the amount of drug delivered to the lung of the animal that was required to treat an RSV infection. Then they would have to study multiples of this dose given over a two week period in both rats and monkeys. The safety work would have to be carried out under the strict principles set out by the FDA for such studies. To do all this, they contracted with Inhale, in Columbus, Ohio. Inhale did a great deal of work with the department of defense on biodefense and had all the systems set up to allow such studies to proceed.

During the RSV development process, Daniel and Iain were forced to work very closely together. They became friends at least in a professional sort of way. But Iain began to miss meetings with no explanation. He would fail to show up for his RSV team meetings where he was the team leader. He would miss meetings with Daniel to review progress and next steps. Daniel asked Iain what the problem was and there was always some reasonable explanation for his absence but never one for the absence of a phone call or other notification. Iain had to pick up his daughter from school because his wife was ill. He himself was ill. His car broke down. Daniel became suspicious and worried. He began to ask questions of Iain's coworkers. They already knew the explanation – Iain drinks – a lot.

Daniel called Iain for a meeting one day to discuss his absences and again asked what was going on. Was Iain having some personal problem? Was there something Daniel could do to help? Iain denied any problem. Daniel confronted him with the belief that he was drinking and not coming to work because he was drunk or passed out. Iain finally admitted that he did have a drinking problem, but denied that it interfered with his work. Daniel pointed out that it did interfere with his work and it interfered with Daniel's work since Daniel was the one to pick up Iain's responsibilities when he didn't show up. Daniel offered to help. Penfrel had an employee-counseling program. They could get him started in Alcoholics Anonymous if he wanted. Iain already went to meetings and had a partner in AA, but was failing anyway. They ended their meeting with no resolution, but Daniel did warn Iain that these unanticipated absences could not continue. They parted in a friendly way but both understood that they stood on the edge.

Daniel did not know what to do. He had treated many alcoholics when he was caring for patients at the VA hospital in Cleveland. He was not naïve when it came to understanding the chances that Iain would turn around. And Daniel needed Iain for his scientific knowledge and, yes, for his leadership skills, which were apparently undiminished in spite of his unexplained absences, at least as far as his team was concerned.

But this changed rapidly as Iain's absences became more frequent and he was no longer able to keep track of the scientific progress being made by the RSV team. Iain's scientific skills were there but unavailable. Here was the man who supported the entire antiviral group at Penfrel during the tragedy of Sasha's murder and the threatened dissolution of antiviral research at Penfrel. Iain was the glue that kept things together while they worked out their future together. He was the person who led the charge for an RSV drug in spite of all the roadblocks encountered by the team – he was the RSV champion within Penfrel. All of this was going down the drain and would not really ever be replaced.

Daniel called Penfrel's human resources department. This step was always a last resort since, although Daniel liked the HR person in charge for infectious disease, he mistrusted the HR group in general. But here, he felt there was no choice. They devised a formal warning letter and a three-month trial period where Iain would have to have no more unexplained absences from team meetings or update meetings with Daniel. At the same time, Daniel had to make plans to replace Iain in case the worst should happen. Predictably, Iain failed within the first month. Daniel would have to let him go at the worst moment for the RSV project. Daniel felt like he was betraying a friend and at the same time, as is frequently the case with alcoholics and their entourage, Daniel felt betrayed by Iain. So did Iain's coworkers who now seemed to be aware of the situation and were supportive of what Daniel was doing.

Daniel asked for Iain's resignation – a much easier process than actually firing him for cause and much better for Iain's future if he was to have one. Iain pleaded only half-heartedly saying that he would change – but both of them knew it was not to be. Daniel had Iain's letter the same day. Iain was gone by the next day.

Within Daniel's group was a good scientist and a strong leader that Daniel had hired just a year earlier – Larry Ryan. Larry looked like a small version of an NFL tight end – someone who could run and block. In fact, he was an avid hockey player and would occasionally show up with a black eye or other obvious bruising from his weekend activities. Larry was a virologist by training who had worked on viruses involved in producing cancer. He had been working at Schering-Plough where he was involved in studying the relationship between how antiviral drugs are dosed, how they behaved in the body and their antiviral activity in animals and people. Daniel hired him to run the pharmacology group that would study how antibiotics and antiviral drugs worked in animals at Penfrel.

Immediately on arrival at Penfrel, Larry was an active and vocal participant in team meetings. He frequently asked questions that kept the teams focused on the goals of the team helping them avoid the distractions of the interesting but not very useful scientific questions that always came up. Daniel was impressed.

Larry was also an accomplished virologist and he was respected, if not always loved, by the scientists in the antiviral group. Iain's group expected that Daniel would appoint the second in charge, a virologist who had been in Sasha's group for many years already, Phil Smith. Phil was a talented virologist, but, in Daniel's opinion, still did not have a good grasp of drug discovery as opposed to academic virology. The team was disappointed when Daniel named Larry Ryan to replace Iain. Several months of low level anxiety and some conflict haunted the group until, finally, Larry earned their respect, was accepted, and things calmed down.

The work on RSV at Inhale progressed rapidly. Inhale called Penfrel and finally reached Larry Ryan after trying to find Iain Cavendish for over a week after Iain's departure. They requested that key members of the RSV team at Penfrel come to the Inhale facility in Columbus, Ohio where the work had been done. They wanted to present their findings in detail – but they sent a summary ahead of time. Daniel himself wanted to go as well as the head of the RSV chemistry group.

Larry made the arrangements and Daniel, Larry and a scientist from the drug safety group headed to Columbus. The Inhale team had first determined how much compound, now called RSV-FI for RSV fusion inhibitor, was delivered to the lungs of the monkeys at which dose. The drug was dosed in general over two hours by aerosol. The dose was increased by increasing the concentration of the drug in the aerosol solution. Inhale had spiked the dosing of some monkeys with radioactive RSV-FI so that they could follow how much actually went to the lower lung where the infection would occur. The radioactive drug could then be measured by a special sort of scan that detected the radioactivity in the lungs of the monkeys.

They then carried out an experiment where they infected several monkeys with RSV, and then treated them with various doses of RSV-FI by aerosol. These were compared to an untreated control group. In this way, they could determine the lowest dose of drug that resulted in a treatment effect demonstrated by decreasing amounts of virus in the lungs. Finally, they treated several groups of monkeys with increasing doses of RSV-FI for a two-week period. There were essentially no toxic effects of the drug for the monkeys. The only problem in these experiments were that at high concentrations of drug, the compound would crystallize out of solution in the lungs, but the crystals appeared to be harmless and would, with time, dissolve anyway. The team was ecstatic and so was Bos when he heard the news. RSV-FI would be the first antiviral from Penfrel ever to be tested in humans.

These results were presented to Penfrel's executive committee and the team had no problem getting support to start clinical trials. Daniel himself, however, still had some misgivings about aerosol delivery and how well caregivers would accept it. Ib Hassad also had concerns on the commercial side not only about whether aerosols would be accepted but also about the fact that RSV was an acute infection. The drug would probably only be needed for a few days. What price could Penfrel obtain? Would the price and the numbers of patients to be treated make Penfrel's investment worthwhile. All these worries remained under the surface as the team prepared to test the drug in human volunteers.

The first trials in humans were carried out at Inhale's facility as well where they had a special unit dedicated to testing aerosol drugs in people. The volunteers received different doses of drug calculated to be below, at the therapeutic dose, about twice that, and about five to ten times the therapeutic dose. The doses were again given over about two hours, and again, the major problem was crystal formation at the highest dose where crystals were formed on skin, eyelashes, hair, and body hair. But there was no coughing and no other real problem for the volunteers even at the highest dose tested.

RSV-FI was about to go to the next step. It would have to be tested in therapy of RSV infections in people to see if it would work to relieve the symptoms of the infection. But how would this be done? Infants under two years of age were the most susceptible to serious infection, but you can't just go from adult volunteers directly into little infants. To test a drug in babies, you have to go very slowly to be sure you won't hurt them somehow. But adults do get RSV infection and occasionally it causes bronchitis in adults. Daniel and Larry got together with the clinicians at Penfrel to try and plan how they would carry out the next trial that would almost certainly have to be in adults with RSV infection. They brought in a number of consultants from various medical schools around the world who studied RSV infection in adults to help with planning.

The first such meeting would be held in New York City. Bos would be there as well. The team planned a day of meetings where the consultants would be presented with all the Penfrel data on RSV-FI. Then they would spend the rest of the day brainstorming how they could go forward. The best suggestion seemed to be to do another volunteer study where volunteers would be infected with RSV and then would be treated with RSV-FI aerosol. They would probably not get symptoms, but the amount of virus they secreted could be measured. RSV-FI should reduce this number quickly. This was not a perfect trial since not all the volunteers would get infected in spite of receiving the virus. Most adults have some level of immunity to RSV. Those that did get infected might clear the virus quickly – but hopefully not as quickly as those receiving RSV-FI. This plan also would require that there be a stock of virus that could actually be used in volunteer studies. As it turned out, there was such a stock at the National Institutes of Health and the volunteer study could be carried out at their facility in Bethesda, Maryland.

That night, the Penfrel RSV-FI team and the consultants sat down to a dinner in a New York restaurant with everyone in a celebratory mood. Ralph George walked in late to the restaurant where everyone had already sat down to dinner in a private dining room. He sat down, tapped his glass with his knife to quiet the room and announced that RSV-FI had run into a slight snag. During standard toxicology testing, Ralph's group had determined that the compound was teratogenic in rats – that is that it caused fetal deformities when given intravenously at low doses to pregnant rats. Even though RSV, when given by aerosol, is not absorbed into the bloodstream well – there was a slight chance that this effect could occur nevertheless. Pandemonium struck. Daniel was furious. Why drop by the dinner meeting with their consultants to drop a piece of news like this when the Penfrel team itself had not had a chance to think about the potential consequences of these data? But this was Ralph. He liked making an entrance and he liked drama. In this case, he was successful on both fronts. Roger Butler noted that Ralph was like a pigeon. "He flies in, shits on everything and takes off again."

Everyone left the dinner in a subdued mood. The Penfrel team and Bos would gather the next day back in Montvale. The clinical group and commercial team would join by videoconference. Daniel knew that RSV-FI was doomed. Even if one could justify exposing desperately ill children to a potentially toxic drug, the healthcare workers who would also be exposed to the aerosol would never accept this risk. In addition, the other drug on the market, ribavirin, which did not work well, was also toxic and delivered by aerosol posing risks for patients and caregivers. Why would you try and market another drug like that? Bos argued strenuously that the drug should be developed for immunocompromised adults. For adults, he argued, the drug could be given with a nebulizer directly through a breathing mask – something you could not use with infants. Infants had to be treated in a mist tent, and the tent was the cause of risk for healthcare workers. Bos argued that the mask delivery system would not pose a risk for caregivers and that the drug could be life saving for immunocompromised adults with RSV infection. But the Penfrel commercial group was not listening. They felt that without the larger numbers of infected infants and without adults and the elderly with RSV respiratory infection, the market would just not be large enough to justify the expense of development. Bos was overruled and there was little Daniel could do. Penfrel's first experience taking an antiviral drug into human trials was an average one – it failed. But there were valuable lessons to be learned from this failure and Daniel would remember them.

CMV, unlike RSV, was very specific for humans and there were no good animal models of CMV infection. You could grow CMV in the test tube – but you would have to go somehow from a drug in the test tube into humans with CMV infection before you would know whether your drug might work or not. This was the opposite of the situation for antibiotics where the animal models of infection were so good that you could even directly predict the human dose based on how the drug worked in mice. The Penfrel experience with their CMV project was worse than that for RSV. It turned out that the entire series of compounds they were working with was chemically unstable. They would disintegrate in water, in saline and in blood. Not only that, but when they fell apart, the products of the reaction were known toxins. This project did not even make it as far as animal testing.

Chapter 9

Viron

In the meantime, the antiviral group, under Larry's leadership, was looking at other viral targets. One of the most tempting was the Hepatitis C virus or HCV. HCV caused an indolent infection of the liver that could lead to cirrhosis and death. It was a common cause of liver failure and those patients were not good candidates for liver transplant since the virus would just go on to infect the transplanted liver. Frequently the transplants would get a virulent infection with the virus probably because of the immunosuppressant drugs needed to prevent the body from rejecting the transplant itself. The virus was spread by blood contact and by other means still not understood. But patients who had received blood transfusions, drug addicts and even health care workers were more likely to get the infection. The disease was actually disappearing because a test for the virus had been developed and blood for transfusion was now screened – so a major source of the infection was being eliminated. Nevertheless, there was a large population worldwide that remained infected and was under threat of progression to serious liver disease, would require liver transplant or would ultimately die from liver failure.

Treatment regimes for HCV were already available. They all relied on using interferon, a substance derived from human cells that was a natural inhibitor of the virus in cells. But, given in the high doses required to have an effect of HCV, it was poorly tolerated with fever, nausea, loss of appetite, muscle aches and other symptoms all being fairly common. The treatment was for 1-2 years and required regular injections. It only worked about 50% of the time. Clearly there was a need for new therapy.

The major problem before the late 1990s was the complete inability to grow the virus and to study it in the test tube. But then there was the invention of the replicon. The virus is basically a strand of nucleic acid that codes for the proteins that the virus needs to replicate and to form its envelope, bud from its host cell and invade other cells. But the strand of nucleic acid is what replicates. The replicon is a genetically engineered version of the viral nucleic acid that allows its replication to be monitored within cells. Infectious virus still cannot be formed, but one can follow the replication of this genetically engineered and non-infectious piece of RNA in human cells in the test tube.

The replicon was the breakthrough that allowed for the discovery and development of new drugs to combat HCV. Penfrel had scientists that had the skills and expertise to use the replicon, but Penfrel had to obtain a license to use it from its inventor – which of course cost money. In addition, Penfrel was behind a number of companies who were already using the replicon to discover new anti-HCV therapies. Larry and his team prepared a white paper proposing a two-pronged approach to HCV at Penfrel. First, they proposed to begin a search for companies already working on HCV with whom they could partner to get a jump-start on finding new HCV drugs that Penfrel could develop and market. Second, they proposed obtaining a license to the replicon, bringing additional expertise in house through hiring an additional two scientists with HCV experience, and to start their own replicon-based HCV discovery program. Daniel was quickly persuaded, especially given the tenuous state of Penfrel's other antiviral programs. He wanted to do something like this before they fell apart completely. Otherwise, there might not be another opportunity.

To do this, it would be up to Daniel and Larry, with some help from Bos Killington, to convince the management at Penfrel that this would be a worthwhile investment. To do that, in turn, Daniel had to enlist the services of the marketing group to examine the potential of the HCV marketplace. But everyone already knew that interferon based therapies, toxic as they were and with only a 50% success rate were already selling over $1 billion globally – so this did not seem like rocket science to Daniel. And, in fact, the marketing folks quickly determined that the market opportunity could be very great with less toxic and more effective therapy. Daniel expected that a drug that specifically targets some vital function of HCV would be at the very least more effective than interferon.

Of course, there were problems with this whole approach. First, the replicon is just a piece of nucleic acid. Could you really go from this to a clinical trial in HCV infected patients? The only animal model that existed at the time was one in chimpanzees. And to do a study in chimps was difficult and expensive even though it could be carried out at a few centers in the US. A study in two or three chimps was still going to be less expensive than one in people with HCV infection. There were, at these centers, a few chimps that were chronically infected with HCV. In the chimp, the HCV does not cause much in the way of liver damage, but is able to establish a chronic infection that could be monitored. Therapy could be tested in these chimps at least in theory. In fact, a few years later, Daniel would do just that, but at another company.

The other problem was that no one had ever done this before. No one had identified an HCV inhibitor using a replicon and then showed that it worked either in chimps or in people. So no one really knew that this could even be done. All innovation starts with this sort of risk. And risk, in a way, is what the pharmaceutical industry is all about. When you know that more than 95% of your scientific ideas don't make it out of the laboratory and 80% of those that make it into early clinical trials will fail, you are already taking risks. But would this be one bridge too far for Penfrel?

Luckily, by the late 1990s when all this was occurring, Brian McKinley was the head of Penfrel R&D. And Brian had come from Penfrel's biotech affiliate, Hemotech. Daniel did not believe that Brian knew much about infectious disease, but he clearly understood innovation and its risks – and he relished these opportunities as long as they were based on strong science and an attractive market. So, in spite of all Daniel's unhappiness with Brian and his management of Penfrel, this was one area where Brian and he would come together in common cause. Larry and Daniel got a green light from Brian and the Penfrel executive committee to carry out their plan as they presented it.

Of course, Larry had already started searching not only for scientists to hire, but also for other companies with whom Penfrel could potentially partner. After the Penfrel go ahead, Larry was able to convene the business development team from Penfrel and to start looking for partner companies in earnest. But he presented them his list of companies in a prioritized order. These companies spanned the globe from the US to Stockholm to Seoul to Japan and back again. Larry and the business development folks began making contacts with the high priority companies on Larry's list. For those where there was some interest on both sides, a trip to meet and review data would be required. Larry's frequent flyer mileage account soared. This work resulted in two companies floating to the top of the list. One was in Stockholm and one was almost next door to the Penfrel facilities outside of Philadelphia. Daniel and a team of Penfrel scientists would visit these two companies. The first was in Stockholm.

Daniel and Larry shared a problem with overseas travel. Neither slept well on airplanes. Penfrel was in the middle of late stage clinical trials of a new sleeping pill called Zomer. The VP of the neurosciences group and a friend of Daniel's had, quite illegally, snuck a small supply of these pills for his friends to try. Zomer was very fast acting and had a very short half-life – which meant that its effects dissipated rapidly. This seemed perfect for air travel. There was one problem – you had to take it on an empty stomach. Eating the airplane meal would not be possible if you were going to take Zomer to sleep. Daniel also knew that Penfrel had just completed a trial where they compared Zomer to Ambien by waking subjects up three hours after they had taken their sleeping pills and gave them a driving simulation test. The Zomer recipients all passed with flying colors, while the Ambien recipients did not do as well. Daniel was anxious to try this new (but still unapproved) pill. The flight to Stockholm was the perfect opportunity. Larry, who sat across the aisle from Daniel, took Ambien to sleep on planes. In this case, they both needed to be rested and alert on arrival because they were going directly from the airport to their meetings. They would then take the evening flight back to Newark.

When they arrived in Stockholm, Daniel was awake, refreshed, hungry and noticed that he had drooled all over his tie. Larry was groggy. "Daniel? I know this sounds stupid, but where are we?"

"What do you mean, where are we?"

"I mean that I see we're on an airplane and we're landing, but I don't remember why we're here. I don't remember going to the airport, I don't remember anything." Larry said in a panicked voice.

"OK, OK, Larry. Take it easy. Calm down. I know what this is. You took Ambien to sleep last night, right?"

"I don't remember."

"Well I do. We talked about it. Ambien has a side effect called retrograde amnesia. That's what you have. It's going to go away soon. Don't worry. In the meantime, I'll be right next to you all the way. When we land, we're going to head directly to a hotel to get showered and change clothes before we head to the company here in Stockholm for our meetings. Do you remember this company in Stockholm?"

"Yeah. OK. I remember. Viral Technologies. Its on the university campus here."

Later, when they met down in the hotel lobby to go to Viral Technologies, Larry was his old self except he was embarrassed and less talkative than usual.

As it turned out, Viral Technologies was trying to discover HCV inhibitors with structures similar to drugs used for HIV. At the end of the meeting, Daniel and Larry were both worried that these drugs might be too toxic. They put the Viral Technologies lower on their list. They decided to move on to the Philadelphia company, Viron.

The President of Viron was an old acquaintance of Daniel's from his academic consulting days, Mike Shannon. Larry had worked in a previous life with Viron's head virologist, Colin Marks. Viron had two projects they wanted to discuss with Penfrel. The first was not HCV, but rather their drug, rhinobrel, that had already completed several small clinical trials for the treatment of the common cold. HCV for them was scientifically exciting, but was of a lower priority for Mike than getting rhinobrel to market.

Mike Shannon was a good scientist, but he was a better salesman. Daniel was able to get Viron back to HCV at least for that meeting, but only after he promised to examine their data on rhinobrel at a later time.

Colin Marks then presented the results of their HCV screening and their chemistry program for HCV all based on the replicon that they had licensed two years earlier. They had made good progress on several different chemical series and had compounds that were not toxic to human cells as far as they could tell, but that did inhibit the HCV replicon at very low concentrations. They also had an understanding of how to make these compounds even more potent while avoiding toxicity with further chemical modifications. Their progress was only limited by the number of chemists working in the company – around seven. Penfrel, if the project was seen as high priority, could put twenty-five chemists on the same project and move it along much more rapidly. Penfrel also had a much more extensive chemical library that Viron did, and might be able to identify additional chemical compounds that were non-toxic HCV inhibitors that might be as good or better than the Viron compounds.

Daniel knew that, because the Viron compounds were not yet in clinical development, a license deal with Viron should not be too expensive. Everyone agreed on additional meetings to look more closely at the precise nature of the chemical compounds Viron already had identified as good HCV inhibitors and to consider what Penfrel could bring to the table. Within six months, they had a deal. Viron would share their replicon with Penfrel. They would work together to obtain improved versions of the replicon. Penfrel chemists would work side by side with the Viron chemists to identify new chemical compounds that might be promising as well as to further optimize the compounds already identified by Viron. Penfrel would pay Viron an upfront licensing fee and assume all expenses for research and development going forward. There would be a joint steering committee to recommend scientific directions and to approve budgets for the collaboration. Daniel would head this committee. There would be milestone payments for any compound that advanced into human trials and these would be substantial. All final decisions were Penfrel's.

Viron agreed to all of Penfrel's conditions except for one sticking point. Daniel insisted that the compounds be checked in chimps before they were tested in people to be sure that they would, in fact, inhibit HCV replication in some living system other than cells in a test tube. Colin Marks was adamant that the chimp model was not relevant to humans and that the additional expense of the study in chimps could not be justified. Since Penfrel would pay anyway – Daniel didn't see the relevance of the financial argument. Once again, Daniel turned to Bos Killington just for a reality check. Bos was quick to agree with Daniel and noted that he was not surprised at Colin's position, but he agreed that animal testing for antiviral efficacy should be performed before testing in humans. Daniel refused to sign off on the deal as constructed by Colin and Viron.

Mike Shannon called Penfrel's CEO, Brian McKinley, to complain. Brian quickly called Daniel. A meeting was arranged between Larry and Daniel and Brian from Penfrel and Mike and Colin from Viron. Both sides presented their arguments, and to Daniel's amazement, Brian agreed with Colin that animal experiments would not be necessary before testing any drugs coming from the Penfrel-Viron collaboration in humans. Daniel could only speculate that maybe Brian's Hemotech background, where Hemotech was a biotech used to taking enormous risks, was behind his thought process.

The collaboration would continue years after Daniel left Penfrel. They would study at least four different HCV drug candidates in humans. The only one that actually worked to reduce HCV infection in humans, the last one tested, had been shown to work in chimps before going into people. Apparently, the Viron-Penfrel collaboration had finally learned that lesson – but it was too little too late. The other drugs tested did nothing to HCV in humans. The one that did work against HCV caused liver toxicity in humans and could not progress beyond its early trials. Penfrel and Viron would never develop or market an anti- HCV drug. There were several lessons from the Viron experience that Daniel would remember in the years to come.

During the time when the Penfrel team was meeting regularly with the Viron team to set up their HCV collaboration, Mike Shannon had many opportunities to try and sell Viron's drug for the common cold. The numbers, on the surface were staggering. Just looking at the US alone, it has a population of over 300 million people. Adults get at around two colds per year and children of school age, especially those under age 12, get about 5 colds per year. Colds last anywhere from five to twenty-one days with an average of nine days. Days lost from work or school are common. Half of all colds are caused by viruses that could be treated with Viron's rhinobrel. Among children and adults with colds, a staggering twenty per-cent will seek treatment from a physician. The bottom line is that if physicians prescribed rhinobrel for only five per-cent of patients calling seeking relief of cold symptoms, the sales numbers in the US alone would be astronomical. These numbers, previously largely unknown to Daniel, got his attention.

Viron had impressive data showing that rhinobrel was a very effective antiviral drug in people. They had carried out a volunteer study where they had infected normal volunteers with a cold virus and then treated half those who became symptomatic with rhinobrel and the other half with a sugar pill. The rhinobrel treated volunteers cleared their symptoms several days more rapidly than those who got sugar pills. There seemed to be no important side effects. No one was allowed to take over the counter medications like Tylenol or antihistamines – a fact noted by Penfrel's commercial group later on. The other more impressive data involved infants with an unusual severe genetic deficiency affecting their immune system such that they were susceptible to infections of all sorts, but especially to viral infections. They would develop a severe and long lasting infection with a close relative of the cold virus, called enterovirus, which was also susceptible to treatment with rhinobrel. This virus would invade their organs and bloodstream and frequently led to death in these unfortunate babies. But rhinobrel treatment cured the infection virtually 100% of the time. The infants did not exactly go on to live normal lives, but at least this particular viral disease could be rapidly and effective treated and for them and their parents, it was nothing short of miraculous. These data, more than anything else, persuaded Daniel that rhinobrel could be an important new opportunity for Penfrel. With permission from Brian McKinley, he invited Mike Shannon and his rhinobrel team to present at the Penfrel executive committee.

The leaders of the scientific, clinical, drug safety and manufacturing groups were all impressed enough with rhinobrel to agree to delve more deeply into the data at Viron. A large team from Penfrel descended on Viron for several days over the next month to dissect and examine every piece of data and all their correspondence with the regulatory agencies looking for anything that would constitute a red flag. Scientifically and clinically, the investigations were all coming up positive. No major problems were identified. But the commercial folks at Penfrel were not so sure.

When the marketing team looked at Viron's commercial projections, they asked – but what if over-the-counter medications like Tylenol and anti-histamines provide symptom relief just like rhinobrel? They dismissed rhinobrel's miraculous activity in immunocompromised infants as medically but not commercially important. This stumbling block of over-the-counter medications eventually caused the entire negotiation to fall apart. Penfrel just could not believe that rhinobrel would be differentiated enough from cheap and easily available products like Tylenol and antihistamines to sell enough to allow Penfrel to recoup its investment. And that was that for Penfrel and rhinobrel.

Daniel, Larry, Mike Shannon and the Viron team were disappointed, but they ended up no better and no worse than before the Penfrel team became interested in rhinobrel – so they continued on their way. Rhinobrel completed two phase III trials with the intention of submitting the data to regulatory agencies requesting approval to market the drug. Unfortunately, the phase III data were disappointing. Everyone at Penfrel got it wrong with the possible exception of the commercial group. In a large population of adults with colds, illness for those receiving sugar pills was about was about eight days and for those receiving rhinobrel, it was seven days. No one was excited about this. But the worst and the most predictable problem had to do with the side effects of rhinobrel. The Penfrel team should have anticipated this one. It turns out that rhinobrel was shown by Viron scientists to induce a liver enzyme involved in the metabolism of drugs – especially hormone-based drugs like birth-control pills. Young women taking birth-control pills had vaginal bleeding in between their regular periods because rhinobrel caused them to metabolize the hormones in the pills more rapidly.

To try and decide whether and how to try again, the Viron team looked at two sub-groups of patients. First, they looked at smokers where the effect of rhinobrel was blunted – there was even less difference with placebo in that group. Then they looked at the use of over-the-counter medications. For those who did not take these medications, there was a greater difference favoring rhinobrel. Viron designed another small trial where they would exclude smokers, exclude women taking birth control pills and prohibit the use of over-the-counter medications. Daniel could not understand this approach since it was so far removed from the real world situation. Apparently, no one else understood this either because the stock price of Viron dropped 80% with the news of the phase III trial results in spite of a simultaneous announcement of plans for a new trial. Mike Shannon retired.

When the rhinobrel news broke, Daniel received a call from a panicking Ralph George, the head of Penfrel's drug safety group. Ralph was a member of the Viron steering committee for the HCV project, and had been involved in the investigations of Viron's rhinobrel program.

"Hi! Daniel! I've got a question for you."

"Hi Ralph. How are you? Shoot."

"I'm good. But I wonder what you thought of Viron's proposed new trial for rhinbrel."

"I think its all over but the shouting. What's the point of studying it in populations that basically don't exist? OK – non-smokers – I get that. But excluding women on birth control pills? Disallowing the use of medications like tylenol and antihistamines? Its just not real. I'm sure it won't even fly with the FDA much less any marketing group in the world. Why are you asking?"

"Well, to be honest, I bought some of their stock."

Daniel tried to hide his gasp of breath. "I'm sorry. Did you just say that you bought their stock?"

"Yes. I know I should not have done that being on the committee that evaluated rhinobrel, but I thought it was going to be a real blockbuster. It sounded great."

Daniel decided that he was better off not knowing this at all. "OK Ralph. I don't know what more I can say. Good luck."

With that they ended their call. Daniel had to catch his breath.

First, Ralph had inside information that other investors did not have. Second, Ralph had a clear conflict of interest in Ralph having Viron stock since he was part of the team that would (and did) decide on whether Penfrel would partner with Viron to develop and market rhinobrel should it ever make it to market. In fact, several years later, Viron would bounce back from its doldrums with a drug for bacterial infection that would sell $500 million per year. Their stock price also bounced back. Daniel often wondered whatever happened to Ralph's investment.

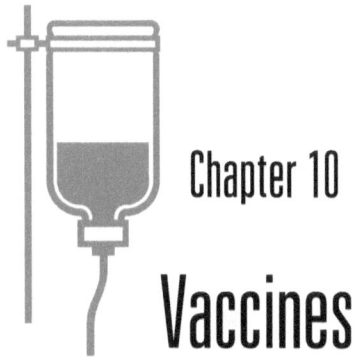

Chapter 10

Vaccines

Penfrel, when it acquired Lederman Pharma, also obtained a very strong vaccine company – Penfrel Vaccines. This subsidiary had its own CEO and had its own profit and loss budget. It was therefore a business separate from the pharmaceutical business in Penfrel. This kind of organization remains common among the large pharmaceutical companies for a variety of reasons. Penfrel Vaccines Research was located in Montvale, New Jersey, just across the campus from Daniel's research group. Daniel was fascinated to find that he knew a number of the researchers in the vaccine group from his interactions as a professor at various infectious diseases meetings. In fact, after his arrival at Penfrel, when Daniel was invited to attend some of the research meetings at Penfrel Vaccines, he was amazed that their research, at least early on, looked much like the research going on in Daniel's group.

Traditionally, vaccine research focused on preventing disease. When Daniel came to Penfrel, though, the Holy Grail was to find vaccines that could actually treat disease. Daniel never thought that this would happen – but as in a number of areas he was wrong. But vaccine therapy would not become a reality until many years later.

Since the research groups had so much in common and since the scientists frequently had complementary skills, Daniel thought that the two groups should merge into one to take advantage of a greater concentration of talent. But David Schubert, the head of research and development at Penfrel Vaccines never agreed that this would be a good idea. Like Daniel, David had come to Penfrel directly from academia. He left Waterloo University in Canada to join the industry having collaborated for many years with the Lederman vaccine group.

There were several reasons for David's belief that vaccines and antibiotics research could never mix, some of which Daniel would only understand several years later. Once the early, most basic research had progressed to a certain point, everything about developing a vaccine was completely different than what occurs during drug development. The safety studies were different. The early clinical trials were completely different and focused more on whether the vaccine might work in people than safety. Safety of vaccines was a concern, but they needed to study many more patients to be assured of safety than were necessary in drug development. Finally, the late stage clinical trials of vaccine were very large, but not terribly difficult to design and carry out compared to the smaller but more challenging trials of antibiotic drugs. While an antibiotic might be approved after study of as few as 1200 patients, a vaccine trial usually comprised tens of thousands of subjects. This makes sense if you are trying to prevent a relatively uncommon disease compared to trying to treat an already established, common infection. In the former case, you have to treat many people to prevent one case of the disease.

When Daniel arrived at Penfrel, Penfrel Vaccines sold an annual flu vaccine and a number of childhood vaccines all of which were generic and none of which made much money for the company. But the most promising vaccine research project was one for a pneumonia vaccine for children. Scientists had known for 100 years that if a person had antibodies against the outer capsule of the bacteria causing pneumonia they would be protected from infection. In fact, in the 1920s and 1930s, before the widespread use of the sulfonamide antibiotics and penicillin, Lederman Pharma raised horses on the land where Daniel now worked in Montvale, New Jersey. These horses were used to raise anti-sera to the bacteria causing pneumonia and the resulting horse sera were sold for therapy. And the serum treatment actually worked – although not nearly as well as antibiotics would work – to treat bacterial pneumonia.

Of course, there were many problems with this approach with allergy to the horse serum being the most important one. The other problem was that there were almost 90 different types of capsule for the bacteria and you would have to raise an antibody for each one separately. Lederman couldn't do that, so they tried to raise antibodies against only the most common and most virulent bacterial strains.

One hundred years later, Merck marketed a vaccine made of purified bacterial capsules from the 20 most common strains causing pneumonia in adults. It is still sold today even though it does not work perfectly well. But Penfrel had found a way to attach bacterial cell capsules to proteins. Done in a certain way, with a certain chemical linker, this combination had the capacity to be a safer and more effective vaccine for pneumonia, especially in children, than the vaccine currently available to prevent pneumonia in adults. And Penfrel thought they had found the secret. The Penfrel marketing team also thought that if they could develop a successful vaccine for children, the regulatory authorities in the developed world would recommend that the vaccine be administered to all children. The rationale for this is that the vaccine would prevent serious infections in children and save many years of productive life, and therefore be very cost effective. Traditionally, vaccines save about ten dollars for every dollar spent for research and then to pay for the vaccine once it comes on the market.

Penfrel Vaccines had a very strong and enthusiastic marketing group. Daniel was constantly struck by the difference between their marketing group and his own within the Penfrel pharmaceutical subsidiary. The Penfrel Vaccines marketers carried out surveys of physicians while the new pneumonia vaccine was still in its early stages of development. They discovered that the pneumonia and meningitis targeted by the new vaccine was rare enough that pediatricians in general pediatric practice would only see maybe one case a year. Therefore, when asked whether they would administer such a vaccine to their young patients routinely, most just shrugged their shoulders. Only 2% thought that they should give their patients such a vaccine. So Penfrel Vaccines, having their own estimates for the number of children that got pneumonia and meningitis every year in the US, partnered with the US Centers for Disease Control in Atlanta to get strong data on exactly how many children would be victims of these diseases and how many might benefit from a good vaccine. This resulted in many publications in scientific journals, in the CDC's own publications, in presentations at meetings and even in good coverage by the lay press. When Penfrel carried out the same survey a year prior to the launch of the new pneumonia vaccine, fully 70% of pediatricians said they thought such a vaccine would be important for their pediatric patients. This was a remarkable education campaign that was, at the same time, a key marketing campaign. An important lesson for Daniel was that sometimes physicians themselves could not recognize a medical need or judge the extent of this need. Sometimes this required a bigger picture view where populations larger than those within a practice of a single physician or even a group of physicians are studied. Daniel asked the vaccine marketing group to make presentations of their work on the pneumonia vaccine to his own marketing group and to his scientists to underline this point.

At their yearly presentations at Penfrel research and development executive committee meetings, Penfrel Vaccines always said that at the price they planned to charge, within two years of the launch of the vaccine, it would sell $1 billion. Daniel remained skeptical – how could these people predict this with such confidence? Later, Daniel understood. If the government authorities recommend the vaccine be given to all children – they buy the vaccine. They buy enough for the entire birth cohort of the US (the western European countries do the same). The vaccine is then administered to all children and the parents or their insurance companies are charged based on a sliding scale. Penfrel Vaccines simply multiplied the price they thought they could charge by the birth cohorts in the developing world and came up with their sales numbers. In fact, their numbers turned out to be a gross underestimate. Penfrel's pneumonia vaccine became the most commercially successful launch of a product in the history of the pharmaceutical industry.

Several other projects at Penfrel Vaccines did not fare nearly as well – but given the success of the pneumonia vaccine – it didn't matter. Penfrel

Vaccines could do no wrong. They would continue to develop and market a number of successful and important vaccines including a follow-up to their pneumonia vaccine. But while Daniel was at Penfrel, there were only two other projects that went into late stage clinical trials.

One project involved a vaccine to prevent rotavirus infection. Rotavirus is a stomach virus that is the most important viral cause of diarrhea in infants worldwide. This was a very exciting project since it would prevent a common cause of hospitalizations of infants for diarrhea and dehydration and even of infant mortality – especially in the developing world. In the US alone, one in every eight infants get symptomatic infection with rotavirus and 2% of all infants end up in the hospital. Rotavirus mostly infects children under 2 years of age.

The problem over the years was that there are many strains of rotavirus, so finding a way to get infants to become immune to all the common strains with one vaccine was difficult. The way Penfrel approached this problem was to make four different viruses. One was derived from a strain that infected rhesus monkeys but which was very similar to one of the common human strains. The others were genetically engineered monkey rotaviruses that contained human rotavirus proteins on their surfaces. All of these viruses could invade human intestinal cells, but because they were specific for their normal monkey hosts, could not damage the human cells. But they could induce an immune response. It was prepared as a freeze-dried preparation of live virus and was delivered orally.

The rotavirus vaccine was approved by the FDA after Penfrel had carried out studies in over 10,000 infants under 33 weeks of age. The vaccine was about 70% effective in preventing severe infection in these infants and there were few side effects. One potential safety problem that was noted during the trials was a rare problem where the small intestine slides in on itself – like a sock as you turn it inside out. This causes intestinal obstruction and can require surgery. In the trials of the vaccine, this side effect, called intussusception, was noted in 0.05% of vaccine recipients and in 0.02% of those who received an oral dose of saline – the placebo recipients. The problem seemed to occur within the first two weeks after vaccination. This difference was not statistically significant, but the FDA and Penfrel established a surveillance system once the vaccine was launched onto the US market where this side effect along with others could be carefully tracked. The CDC was quick to recommend the vaccine for all children and Penfrel thought they had another blockbuster vaccine.

But, the surveillance system worked in the sense that Penfrel was able to collect data on most of the vaccine recipients within the first year the vaccine was on the market. These data supported the information available from the original trials suggesting that the risk of this side effect was twice as high in the vaccine recipients as in those who did not receive vaccine. Several large meetings were held with the CDC, the FDA and scientists and clinicians from around the world to review the data and to decide whether this increased risk was outweighed by the benefit from the vaccine. Even after reviewing all the data, it remained controversial whether the increase in intussusception was real or a statistical fluke since the numbers were still small. And there was no real consensus that the incidence of this problem outweighed the overall benefit of the vaccine. In spite of the controversy on these scientific points, Penfrel itself finally decided that the vaccine had been so tainted by the bad press around this side effect that it would be difficult to sell in any case. It was also not clear that the CDC would continue to recommend the vaccine for all infants. Penfrel withdrew the vaccine from the market after just one year of use. No one ever understood what it was about the vaccine that might have led to intussusception. And whether this was a real side effect of the vaccine remains controversial. Penfrel's rotavirus vaccine was never licensed outside the US. There would not be another rotavirus vaccine available for children for another six years and these vaccines would come under microscopic scrutiny especially for the potential problem of intussusception. The newer rotavirus vaccines apparently are more safe and do work well but no one knows why they do not cause intussusepcion and the Penfrel vaccine apparently did.

The most controversial project in Penfrel Vaccines was a project where they collaborated with the Neurosciences department in the pharmaceuticals business to develop a vaccine to prevent or at least halt the progression of Alzheimer's disease. If you examine the brains of patients dying with Alzheimer's, you routinely find tangles of protein fibers in key locations in the frontal lobes of their brains. These abnormal protein fibers are thought to contribute to the establishment and progression of the disease. Surprisingly, there is a mouse model of Alzheimers where the mice develop abnormal behavior, stop eating and eventually succumb after a few months. They develop protein fiber tangles similar to those seen in humans. Penfrel had a collaboration with a group of scientists at an Ivy League university who had been able to raise antibodies to these aberrant protein fibers in mice. They could show that the vaccine they used would prevent the establishment or the progression of the disease in the mouse model and that the mice who received the vaccine suffered no ill effects. The scientists at Penfrel Vaccines worked with the Penfrel neurosciences group and with the university scientists to develop a similar vaccine for use in humans.

Daniel and other scientists at Penfrel immediately identified the risk that might be associated with such a vaccine. The abnormal protein tangles derive originally from normal proteins in the brain. Would the immune response to the abnormal protein be specific enough to be safe? Or would antibodies also be formed to normal brain tissue and lead to disease in those receiving the vaccine? To find out, the Penfrel scientists proposed testing the vaccine by starting with a very low dose and increasing the dose slowly in patients already suffering from Alzheimer's at a very early stage and were willing to participate in the trial. Clearly, they would have to still be mentally competent such that they understood what they were signing up for. Given the desperation of many Alzheimer's sufferers, it was not hard to find willing participants.

At the time of this trial, Daniel was assigned to participate in the ongoing safety committee at Penfrel. The safety committee dealt with potential and real and emerging safety problems in all aspects of Penfrel's drug and vaccine development both in animals and in human trials. They met once per month under the leadership of Ralph George, the committee chairman. One night, around midnight, Daniel was awakened by a call from Ralph. He announced that there would be an emergency safety committee meeting at 7 am the next day. Ralph himself was calling from London, so for him it was already 5 am. The meeting would be by videoconference and Daniel would join from the Montvale facility.

After calling the meeting to order, Ralph asked the clinician from Penfrel Vaccines to explain the reason for the meeting. So far, about 60 patients had been enrolled in the Alzheimer vaccine trial. Twenty patients each had received either placebo or a low dose or high dose of the vaccine. About six subjects, five from the high dose group and one from the low dose group had developed symptoms of encephalitis. Several were hospitalized and one or two were comatose. Everyone was still alive. The vaccine group then went on to explain that this was a risk that they had anticipated given the way the vaccine was prepared and that this risk was noted in the informed consent that the patients had signed when they agreed to participate in the trial. Daniel was shocked since this was never discussed at any of the Penfrel meetings where the vaccine was presented on numerous occasions. At the suggestion of the vaccine clinicians, the safety committee agreed that the trial should be halted until they could figure out if this was a coincidence caused by an epidemic viral encephalitis or if this might be a specific effect of the vaccine. Daniel knew that there was no encephalitis virus circulating in England at the time and he knew then why Ralph was in London. The trial was enrolling patients in England. Only one more patient would become ill from the vaccine after the trial was halted.

There were never any more trials involving the vaccine. And all the patients recovered from their encephalitis. These patients were followed very closely for years after participating in the trial. It turns out that the vaccine recipients, including those that suffered from encephalitis, had a temporary or even permanent halt in the progression of their dementia suggesting that in spite of the safety risk, the approach might be a valid one. But no one has been able to develop either another vaccine or a successful therapy for Alzheimer's based on what was found during the Penfrel vaccine trial.

When Penfrel eventually abandoned its antibiotics and antiviral programs firing most of those scientists, it maintained its vaccine subsidiary and even strengthened it. A few of the antibiotic scientists were even able to find jobs in Penfrel's vaccine group.

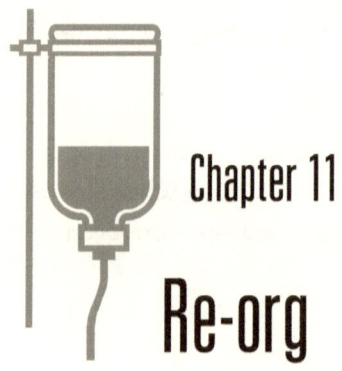

Chapter 11

Re-org

When Daniel arrived at Penfrel in 1996, they already owned majority interest in a biotech company in Cambridge, Mass. called Hemotech. Late in 1997, Penfrel purchased the rest of the equity in Hemotech. The CEO of Hemotech was golfing partners with the CEO of Penfrel at the time. Both owned homes on Nantucket and interacted frequently when there. Penfrel's CEO made the decision to make the Hemotech CEO the President of Research and Development at Penfrel. During Daniel's six-year tenure, Penfrel would undergo three significant reorganizations, but none were as traumatic as this one. This decision meant that the two principal players in Daniel's life at Penfrel over the prior year, Alan Smith and George Finkel, the President of Penfrel Research and Development and Alan Smith's boss when Daniel was hired, would both be relieved of their responsibilities.

Daniel and all of the VPs at Penfrel admired George intensely. George was a physician scientist who had worked on the metabolism and synthesis of cholesterol at the National Institutes of Health. He won the Lasker Prize in medicine for his work – one notch down from the Nobel. Daniel had interviewed with George when he came to Princeton to interview with Alan Smith. George was a wonderful human being as well and Daniel would call him once in a while just to chat about how things were going or were not going.

George would stay on for several more years as a "special advisor" to the Penfrel CEO. At the time, George was about 60 years old. He would be struck down with cancer within a few years without ever having been able to retire to spend time with his children and grandchildren.

Alan was simply let go. He went on to start and then either sell or close down several biotechs and he continued his marathon running – all from a large and gorgeous house in Indainapolis, Indiana.

This reorganization also meant that the clinical development folks with whom Daniel had worked to build relationships over the previous year would no longer be involved in infectious diseases projects. Instead, Hemotech people would fill all these positions. The new head of Research and Development, Brian McKinely, called a meeting of all the vice presidents in charge of the Penfrel discovery areas – oncology (cancer research), women's health, neurosciences, medicinal chemistry, and others. The meeting took place in a hotel not far from Penfrel's corporate headquarters in Madison, New Jersey. Brian had also invited a few of the senior scientists from Hemotech including his head of discovery there, Ron Gordon (Alan Smith's counterpart). He first announced that Alan would be leaving. The Hemotech head of discovery, Ron, would replace him. Ron had spent his entire career after receiving his PhD from Harvard University at Hemotech. He knew nothing about the kind of drug discovery (so called small molecules or chemicals) done at Penfrel. He did know a fair amount about the discovery and manufacture of protein drugs – but that would not help him lead Penfrel. Ron knew nothing really about antibiotics, antivirals or infectious diseases and neither did Brian. Brian then started to draw an organizational chart outlining his vision of the new Penfrel R&D. By the end of his presentation, which he carried out on a whiteboard, there were so many arrows going in so many directions that Daniel no longer knew who was doing what for whom.

The Penfrel VPs spent the rest of the afternoon in a nearby bar. They could barely speak they were so depressed. It was clear to them that people who had no idea what they did or how they did it were about to become their leaders. The VPs were especially worried about Ron who looked like a deer in the headlights during Brian's presentation. That evening, there was a "get acquainted" dinner. Alan Smith was there and loudly tried to cajole Brian into letting him stay on. That was embarrassing and sad and led nowhere.

Daniel called Sally with the news. Her advice was – keep your head down and keep doing what you're doing. Daniel went back to Montvale and announced the upcoming management changes to his team. Depressed would be an understatement of the response from everyone. But, everyone had work to do and everyone still had jobs. Daniel passed on Sally's advice.

In the meantime, Brian McKinley started getting active in his new job. One good thing he did was to establish a matrixed organization called the Therapeutic Area. This organization would be headed by the research VP – Daniel for infectious diseases – and a colleague from the marketing group.

Daniel wanted to work with his friend of many years, Roger Butler. But Mark Mann, the head of the new products marketing group at Penfrel had other ideas. Mark and Daniel had butted heads on several previous occasions, the most divisive being their argument over the potential commercial value of mycin. But the decision on who was to lead the marketing component of the therapeutic area was Mark's, not Daniel's responsibility. Mark appointed Hassan Rashid. Hassan had been the head of Penfrel's affiliate in Dubai and knew little about antibiotic research. He did have an understanding of how to sell antibiotics since Penfrel's peracillin was his best selling drug. Of course, selling antibiotics in Dubai, at least to Daniel, was probably not the best preparation for evaluating new antibiotics and antivirals coming out of a research organization.

In a way, since Roger was really the only person in new products marketing who understood the antibiotics market, Hassan would be obligated to ask for Roger's help in any case. This would turn out to be good for Daniel, but not so good for Roger. Mark Mann was determined to undermine Roger at every turn. Daniel always thought that this was related to Mark's ongoing battles with Daniel over the valuation of infectious disease projects.

Together, Hassan and Daniel would build a team from every functional area within the company including clinical development, manufacturing, drug safety and everything else required to develop a drug. Of course, the only people actually directly under Daniel's supervision were those in his own research group. The various functional area heads supervised all the others. But Daniel could rule by fear. If he was not getting cooperation from a functional area he would first call the head of that area and try and cajole further cooperation. If that didn't work and Daniel still felt he was right, he would call Brian McKinley – something that everyone wanted to avoid. For the most part this worked. Daniel actually had some control over everything from the earliest stages of drug discovery all the way through sales strategy for marketed products. Not only did this work, but for Daniel, the therapeutic area became a post-doctoral course in the pharmaceutical industry. He would have to learn all of the functional areas in enough detail that he could actually interact with them in an authoritative manner. When Daniel first arrived at Penfrel, he thought it took six months to find the bathrooms. As therapeutic area head, the learning curve was even steeper.

On Daniel's arrival at Penfrel, Sally would call work occasionally to discuss the move, the house and a variety of other things. At first, she had difficulty understanding Daniel's two secretaries who had strong Brookyln and Bronx accents. The secretary who worked most closely with Daniel was Rose who was from Brooklyn originally. Rose had never traveled outside New York State and had barely ventured out of the New York City metropolitan area. As Daniel traveled around to the various Penfrel facilities to meet everyone and to see what went on at these different sites, Rose was given the task of arranging all the travel. This was a mistake at least at first. Daniel would find himself in his car on the way back from the 250-mile drive to Penfrel's clinical research center outside Philadelphia when he was reminded that he had a meeting back in Montvale within the hour. After several such episodes where Daniel was embarrassed and either late or absent, he determined to have a serious chat with Rose. But Rose was charming, adorable and considerate. Daniel found that he couldn't even lose his temper with her. So he began to personally supervise the travel arrangements she was making to be sure that they fell somewhere within the realm of the possible. After the first six months of this, Rose was able to carry this task independently. Daniel called Tom Franklin at one point since Rose had worked for Tom for five or six years. "Oh yeah." Tom said. "Rose could never do that. I always made my own travel arrangements. Why do you ask?"

Daniel's office and the research group he supervised directly was in Montvale, New Jersey. He had a five-mile commute to work and in good weather he rode his bike back and forth. The establishment of the therapeutic area would expand Daniel's responsibilities to cover people working at Penfrel's clinical research and marketing facilities in Collegeville, Pennsylvania a three-hour drive from Montvale. He would also have to visit the European clinical development offices in Paris on a regular basis. Essentially this meant that Daniel was on the road about 30% of his time. This was a complete change from the days where he was able to concentrate his responsibilities in Montvale. Sure – he would have to travel once a month or so on a regular basis, but nothing like 30%. This also meant that he became more and more pulled out of the day-to-day science that he so loved. It also meant that he would be constantly exposed to his upper management – something he would have preferred to avoid. But Daniel would come to enjoy his expanded responsibilities, if not the travel, and he would ultimately want to expand them further. That next step would never be possible at Penfrel, especially for someone in infectious diseases like Daniel.

Penfrel was already selling an antibiotic called peracillin that had been developed by Lederman Pharma under Tom Franklin. In fact, Daniel had worked on peracillin when he was at the university in Cleveland. It was selling under $400 million when Daniel arrived at Penfrel in 1996. On arrival, Daniel met with the marketing directors for peracillin.

Daniel, while still in Cleveland, had carried out research showing that peracillin was less likely to select for resistance during treatment of patients than another class of antibiotics, the cephalosporins. Daniel suggested that Penfrel use this information in their marketing campaigns in order to convince physicians and hospitals to replace the cephalosporins with peracillin. The campaign was successful and sales started to climb rapidly. But peracillin was manufactured at two facilities at Penfrel – one in Puerto Rico and the other in Catagna, Italy – just under the volcano, Mount Etna. These were the only two facilities able to manufacture the drug for Penfrel. Peracillin was a pencillin antibiotic. As such, by law, it had to have its own, isolated manufacturing facility. There could be absolutely no contamination of other products with a penicillin because some patients were so allergic to penicillin that the slightest amount could be lethal. Both of these manufacturing plants had constant problems. Puerto Rico was an old facility that needed updating, but Penfrel was reluctant to invest in infrastructure, especially in Puerto Rico. Contamination either with particulates or even with microbes was a recurring problem. Mount Etna would erupt on occasion forcing the facility there to close for various periods of time. The sales people were constantly trying to balance a limited supply of peracillin with increasing demand. Frequently, they would have to tell hospitals that their shipments would be delayed by a month or more while manufacturing caught up with demand. This was not good for sales. This meant that the climb in sales always lagged behind predictions. It was not until years after Daniel left Penfrel and near the expiration of peracillin's patent that its peak sales of over $1 billion dollars occurred.

Penfrel's lack of investment in their manufacturing infrastructure came back to bite them hard a few years after Daniel's arrival. One day a company-wide large screen videoconference where attendance was mandatory was announced. Richard and others asked Daniel what was going – but Daniel had no idea. The videoconference was run by Brian McKinely the head of Penfrel research and development, the head of manufacturing, Ib Hassad, the Penfrel corporate counsel, Harold Holtzman and Penfrel's CEO at the time, Barry Boswell himself. Of these folks, Daniel only really knew Brian and Ib. He had met the others at various company meetings and social occasions, but did not know them well. The meeting was called to order my Mr. Holtzman. He announced, backed up by powerpoint slides with large lettering, that Penfrel had just entered into a compliance agreement with the FDA. The room in Montvale burst into spontaneous gasps and quiet outbursts of conversation. Daniel had to ask for quiet just so he could hear the rest of the announcement. Daniel had no idea what this might mean, but apparently others in the room already had a good idea.

What had happened was that the FDA was regularly inspecting Penfrel's manufacturing facilities in the US and even some of them outside the US. Many had recurring deficiencies noted for several years in a row by the FDA inspectors but which Penfrel had not addressed. Ib had pleaded with Penfrel's CEO and others for years for funding to deal with these deficiencies, but the company always had other priorities – usually involving reporting profits to Wall Street and increasing the company's stock value. In addition, Ib was only responsible for the manufacturing facilities involved in making compounds in development – and even then he did not get involved in vaccines which was a separate business within Penfrel. Marketed compound manufacturing fell to the commercial group within Penfrel. In order for Penfrel to take a systematic approach to manufacturing, the research manufacturers, the commercial manufacturers and the vaccine manufacturers would have all had to get together to develop an overall plan. This was like asking Homeland Security, the CIA, FDA and NSA to all coordinate with each other. Not only did Penfrel never take this approach, Daniel was not sure that they ever even considered it.

The FDA had written Penfrel a warning letter one year earlier. Six months prior to the videoconference, the FDA threatened Penfrel with criminal action and the possibility of charging Penfrel officers with felonies for lack of compliance with FDA regulations in their manufacturing facilities over a number of years. This last letter finally got Penfrel's attention. Jail time for Barry Boswell was apparently not in his retirement plan. Harold Holtzman and his team of lawyers, along with the department of regulatory affairs at Penfrel started a negotiation with the FDA to avoid a criminal complaint and to either close manufacturing facilities or bring them into line with FDA regulations. But the process of fixing these facilities would take several years and would have to be carried out under the supervision of FDA inspectors. Penfrel was also obligated to pay a fine of about $200 million – the largest ever levied against a pharmaceutical company at that time. But Penfrel would not have to admit wrongdoing.

Brian McKinley then went through some of the other steps the company would take. Ib would head the compliance effort across all of Penfrel including commercial and vaccine manufacturing. He would hire a number of manufacturing engineers and consultants to help decide which facilities would remain open and be brought into compliance and which would be closed. Ib then announced that the company had already decided to close three facilities in the US that they felt would be too expensive to bring into compliance and whose work could readily be outsourced to other Penfrel facilities or outside contractors. Two of these made vaccines for Penfrel and one made an antibiotic as well as the starting material for teracil. This would mean the loss of several thousand manufacturing jobs in the US but at the same time, a couple of hundred engineers, scientists and consultants would be hired.

One result of all this was that decisions around manufacturing would be slowed since everyone in that department was about to be overwhelmed with compliance related tasks. Another result was that Barry Boswell, who said not word one during the videoconference, would retire within the year. Daniel was incensed. As much as Daniel thought the FDA anti-infectives group had gone crazy, here he had to agree with the FDA. Was this a way to run a company? He also worried that the teracil project would be affected – but this turned out not to be the case. Daniel also wondered what would happen to peracillin manufacture in Puerto Rico and Catagna – but these sites were updated and continued to manufacture peracillin.

Penfrel, like most large pharmaceutical companies, made its investment decisions based on the concept of net present value. For any drug development candidate, an assessment of the net present value had to be made and some minimal value had to be achieved. The net present value calculates all of the anticipated revenues for a drug over the lifetime of its patent protection (before becoming generic) minus all the anticipated costs for the product including research and development, marketing and others. All this is corrected for inflation at a standard (but wildly conservative) ten per-cent per year. Of course, for the early stages of development where the investment would be minimal, the sales estimates were wild speculations. Only as the drug progressed could sales estimates be honed – but even then, the predictions were certainly not good science but remained, at best, educated guesses. For antibiotics, the year of peak sales, as was the case for peracillin, tends to occur late. But late revenues are so heavily discounted in the net present value calculation as to be virtually worthless. As such, most such calculations for antibiotics result in figures of around $100 million whereas those for drugs that are taken over the lifetime of patients, like the statins, are 10s of billions of dollars. So in any competitive portfolio review process in the pharmaceutical industry, antibiotics are destined to lose. There was no exception to this at Penfrel. It was for this reason that the marketing group estimated a peak year sales of $1 billion for teracil when that number was almost surely a gross overestimate. Sticking with reality would have doomed the drug to portfolio review oblivion. Daniel and Roger Butler may have been correct in their own subterranean sales estimate for teracil, but it was better that it remain underground.

Brian Mckinley's rule at Penfrel would come to an abrupt end. With the resignation of Barry Boswell following the manufacturing fiasco, a new CEO was chosen from among the ranks of Penfrel management – Hank Wallace. Daniel didn't know Hank, but several of the other VPs did and they knew that Hank had been shocked by Barry's move to appoint the Hemotech management to such dominant positions within Penfrel. They began to cautiously approach Hank with their complaints about Brian and his inept management of research and development at Penfrel. Daniel, even though he agreed with their assessment of Brian, especially following the confrontation with Viron where Brian overruled what Daniel thought was his own very reasonable position, kept his head down and watched from the sidelines.

As is often the case, you should be careful what you wish for. Hank quickly replaced Brian with Sam Stern. Sam came from another large pharmaceutical company, Gorman Ltd, and had a reputation as a ruthless manager with poor people skills. Daniel knew a number of people in the Gorman infectious disease group from the days when he consulted for SKB (many of the ex-SKB scientists had been hired by Gorman). He called them right after the announcement was made at Penfrel. They recounted how Sam and the head of their group used to announce to the infectious disease group at the beginning of every year that not all of them would be there the next year. In other words, given the rating system in place at Sam's company, there was an obligation to give a failing assessment to about two per-cent of the employees every year. When you spread that two per-cent across all of research and development that would mean that some areas would have more firings than others. The employees felt under constant threat since the measures that were used for evaluation were always somewhat arbitrary and completely depended on the whims of your manager.

No matter how hard the human resources people tried to make these evaluations objective and to remove any bias, they could never succeed. Human interactions remain what they are and emotions play their role in spite of all efforts to the contrary. Apparently, Sam relished the idea of holding this possibility over his employees as a way of stirring them to greater accomplishments. Daniel was shocked on hearing this since his management style was diametrically opposed to Sam Stern's approach. Daniel actually believed that employees could improve when placed on a performance plan when they were underperforming and this had been true in several, but admittedly not all, cases. On hearing of Sam's aggressive approach to personnel management from Sam's previous company, Daniel once again decided that Sally's head down approach was the best one.

Sam took a tour of all the Penfrel sites meeting with all the VPs heading Penfrel's therapeutic areas. When he arrived in Daniel's office, he was all openness and was positive about all the programs at Penfrel. He seemed genuinely excited about his new job and seemed warm and open. He told Daniel that his door would always be open. He also told Daniel that he had not yet made any decisions about what would happen with Ron Gordon – ostensibly Daniel's boss. Ron had shown himself to be just as Daniel and the other VPs had feared. He was quiet, shy, virtually incapable of decision, and clearly did not understand what the Penfrel therapeutic areas working on small molecules were doing. Having come from the Hemotech side, he was very well versed in the discovery and development of protein drugs, but not small molecule drugs. He certainly knew nothing at all about antibiotics or antivirals. In addition, he had appointed a number of other Hemotech scientists to key positions within Penfrel and Daniel felt that there was a clear bias toward supporting them to the detriment of those on the Penfrel side of things. When Sam Stern raised the possibility that Ron's days might be numbered, Daniel's ears perked. Sam then said that he had made no decision regarding Ron and he noted that if there was a problem with Ron, it would be up to those in Daniel's position to help Ron overcome his deficiencies. At the end of the discussion, Daniel was left not knowing what to think. He called Robert Stein across the way to talk about it. Bob said, "Keep your head down." That's exactly what Daniel tried to do.

But there was no avoiding confrontations with Ron Gordon nor was there a way to avoid discussions with Sam Stern. Daniel would be asked to meet with Sam on a regular basis. Sam started work in his Collegeville office around 5 or 6 am. His first appointments were usually at 6:30. Sam acquired the reputation of having a very short fuse. Daniel supposed that when you work on just a few hours sleep every night this might be understandable. But he also knew from his conversation with his colleagues at Gorman, that there he was known as Yosemite Sam both for his temper and his ability to speak without thinking at times (shoot from the hip). The Penfrel VPs quickly picked up this pseudonym when talking about Sam among themselves.

Daniel was frequently at Sam's office by 7. This meant he would have to travel down the night before, heading back to Montvale the next day. Of course, he always tried to combine his visits to Sam with other work in Collegeville. His discussion with Sam mostly revolved around Ron Gordon and his inability to make decisions. The problem was that Ron could not prioritize anything. So everything that was submitted to him for approval, he approved. But all of these requests could not be absorbed within Ron's budget. So Sam would call Ron and ask for a prioritized list of what he wanted. But Ron could never provide that. So Sam started trying to prioritize things for Ron and this led to the regular meetings with the Penfrel VPs including Daniel. To Sam's credit, he tried everything. He tried team-building exercises. He tried counseling sessions both for the VPs and for Ron. But, in the end, Ron had lost all respect of the VPs and this was never going to be repaired.

Sam finally let Ron go and replaced him with a colleague from Gorman, Martin Moran. He, like Sam, worked on heart disease and had no understanding of infectious disease. Daniel's first meeting with Marty was enough to convince him that his head down stance was not going to change anytime soon. Daniel did everything possible to avoid meetings with either Marty or Sam.

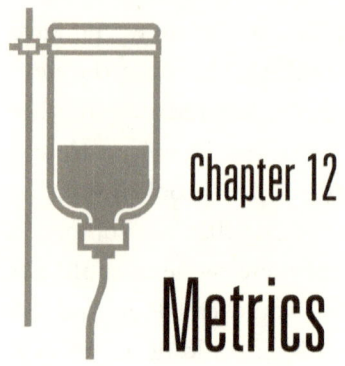

Chapter 12

Metrics

Most large organizations have systems that allow managers to evaluate employees and to provide rewards for good performance and incentives for those who are underperforming. Daniel and his group at Penfrel had a system focused on what they could realistically promise plus some "stretch" goals that would indicated whether they were able to go beyond what seemed possible when putting these things together. Daniel reminded everyone, including himself and his own boss, that failure is the rule. We need to stick to the science, push our project forward, and recognize when we've hit the brick wall so we can move on to something new and hopefully better. This system was working under Daniel, his boss, Alan Smith and Alan's boss, George Finkel. Daniel's group was succeeding in their goal of delivering promising antibiotics and antivirals for clinical development. The fact is, that most people choosing to do research in the pharmaceutical industry have that as their personal goal to start with. They truly want to deliver useful drugs to patients and physicians. Sure – they need focus and guidance at times – but the motivation is already there at least for the most part.

During the 1990s and at the turn of the century, "metrics" was all the rage in large organizations. How do you motivate people to get things done that matter for the corporation? How do you measure whether they are succeeding? What is the best way to distribute rewards on the one hand and incentivizing punitive measures on the other?

When Yosemite Sam arrived, he introduced the Gorman system of "metrics" to Penfrel. This would mean that, like the Gorman system, there would be a small number of employees who would receive "excellent" performance evaluations, but an equal number who would receive failing grades – by dictate. Performance evaluations for employees at these two extremes became a negotiation among all of Penfrel's senior management every year. The bonus budget was a pie for the research and development group of 5000 employees. If one group had two per-cent excellent employees but one per-cent underperforming employees, someone else would have to make up the difference or that manager would have to find the other one per-cent somehow.

Daniel, by chance, was able to balance out the evaluations within his own group most of his remaining years at Penfrel, but this codified distribution of rewards and punishments was devastating to morale among the scientists just like it had been at Gorman. It arbitrarily divorced rewards and punishment from performance and in Daniel's opinion was counterproductive. He did not make himself popular with Sam when he expressed his opinion.

Perhaps the greatest folly of the metrics was the one imposed on the research groups. Sam insisted that the Penfrel research scientists deliver no less than ten compounds every year into clinical development. That number gradually increased to fifteen before Daniel left the company. Sam made this goal also part of the performance evaluations for the heads of manufacturing, drug safety and clinical development. So the entire research group shared this delivery goal. The result of this was that the research group met their delivery goal every year, but the failure rate for these compounds increased compared to the system that worked without these metrics. The failure rate for compounds entering into clinical trials in patients was 80% across the industry. Penfrel's failure rate shot up from the average to 90% as a result of Sam's metrics. Yet another goal devised by Sam was to reduce the failure rate of compounds entering clinical trials. His delivery metric undermined his own goal of improved success rates. In fact, among the VPs who were responsible for delivering these compounds for clinical development, this metric became a standing joke. Everyone tried desperately to provide a compound from their respective shops, but everyone also knew that this metric would do nothing but increase the chance of failure since things were being pushed through the system that probably should not have gone forward. At the end of the day, Penfrel was spending more money trying to develop a greater number of compounds of inferior quality rather than focusing on a smaller number of compounds with a greater chance of success. This metric was a complete failure.

The entire system of metrics introduced by Sam served only to sour the scientists' relationship with Sam and the Penfrel management and hurt the company's bottom line in the long run. Yet, as the number of compounds in Penfrel's development pipeline increased, Sam went around the world giving speeches to other companies about the success of his metrics.

Daniel spent a good bit of time trying to think about ways to formulate metrics that would encourage the development of successful compounds. The way Sam implemented his metric for delivery of compounds tied all the research heads together at the ankles. What if you made this a goal for the discovery heads but not for the development heads like manufacturing, drug safety and clinical? In Daniel's opinion, this would just start wars between the discovery groups and the development groups who would be at constant loggerheads about which compounds could and could not go forward. Daniel eventually gave up on the idea of compound delivery as any sort of metric.

Penfrel offered its employees stock options. Daniel initially thought that this should be a motivating factor for drug hunters. But it turns out that so many things can effect the stock price that are simply out of the control of scientists, that the stock option does not provide a specific motivation to scientists to do their specific jobs. In fact, after the first few years of Daniel's tenure, the FDA compliance decree, the failure of a compound in late stage development and the withdrawal of a marketed compound for unexpected serious side effects all caused the stock price to drop and caused Daniel's stock options to lose value to the point where most were under water. So it was never clear to Daniel that stock options were the answer.

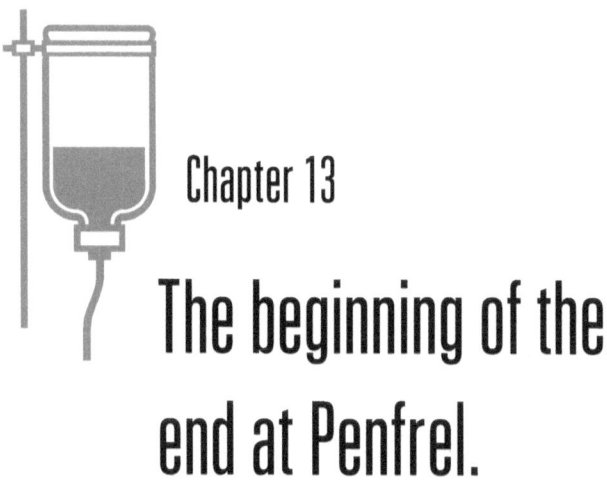

Chapter 13

The beginning of the end at Penfrel.

At Penfrel, Daniel's reputation soared. His successful interaction with the FDA around teracil became the high point of his career at Penfrel. So Daniel was shocked to get a phone call from Sam Stern in the fall of 2001, just a few months after the FDA meetings on teracil. Sam explained that the executive committee and the CEO of Penfrel wanted to review the current status of infectious disease research and development in the company. In the background of several other companies abandoning antibiotic research, Daniel knew that this could be Penfrel's way of getting out. He lost his temper – never a good thing when talking to Yosemite Sam. "If you want out of antibiotics – just tell me. There is no need for a meeting if you've already made your decision!" Sam insisted that there was no such intent (lying through his teeth) and that this was just a review. "If it's just a review, when will the other therapeutic areas like cancer, neurosciences, inflammation and the others be reviewed? Are we the first or the only ones on the block?"

"I'll call you back" Sam replied.

Daniel and Sally had been in Boston at the time. Daniel had meetings at the Penfrel affiliate there, Hemotech, and they had combined the trip with a visit to family in Cambridge. They drove home slowly under a deep blue sky and magnificent fall foliage staying overnight in western

Connecticut before heading back to their home in New Jersey. They spent the drive discussing possibilities for Daniel's next career move since he believed that this would be the end of his sojourn at Penfrel.

One of the career moves they considered was a move back to academia. Sally desperately missed their friends and their house in Cleveland. But when Daniel thought about this, he realized that at his level (he had been a Professor of Medicine) in order to return, he would have to have grant funding. Why should a university medical school hire a research professor who had no research funding when there were lots of job applicants who did have funding? No, Daniel would have to find another job in industry. But how? Where?

Not long after their return to New Jersey and work at Penfrel, Daniel did receive a call from Sam. He said that he had decided that Daniel was right and that the reviews should encompass all therapeutic areas. He asked when would be a good time for Daniel to present infectious disease. He said that the CEO of Penfrel himself, Hank Wallace, would be there as well as all the business and science department heads. Hank Wallace had climbed through the ranks of the Penfrel sales and marketing group to become the CEO, following the ignominious departure of Barry Boswell. Barry had been responsible for the manufacturing compliance fiasco with FDA as well as the replacement of a strong research and development leadership at Penfrel with a much less talented group recruited from Hemotech. Daniel had never presented to the CEO before and had only met him briefly at various Penfrel functions. Daniel had the impression that Hank was a reasonable man. But Daniel had an abiding mistrust of Yosemite Sam. Sam gave Daniel a time frame running from March to June of the next year. This would give all the therapeutic areas time to prepare. They agreed on a date in May – eight months later. Daniel was surprised that the tension of their earlier conversation seemed to have disappeared, but he remained extremely suspicious of Sam and the Penfrel research management. At least he could put off his worries about losing his job for a while.

Within a few weeks, Sam sent out an outline showing in detail what the committee wanted to review. This included a review of sales and marketing of already marketed products, late stage projects (in late clinical trials) complete with sales projections, and earlier stage projects with sales projections. The committee also wanted to see a forward-looking strategy noting which areas the therapeutic area was targeting for new products either through internal efforts or through acquisitions from outside Penfrel. A half-day – 8 am until noon – would be dedicated to the review.

Preparations for the review consumed most of the next eight months. Daniel and Hassan worked furiously to find help in getting the required sales projections for the earlier projects together. They finally settled on an outside contractor and Hassan (not Roger Butler) was assigned to supervise their work. The US and global managers for peracillin, their marketed antibiotic, would handle the presentations for peracillin market projections and sales strategy for the next few years. Daniel had the job of presenting the research and development strategy for all of the late stage and earlier projects both in the lab and in the clinic. He would coordinate with Hassan and the team putting together the sales projections for these projects. Finally, Daniel and his team would work on presenting their strategy for the future of infectious diseases at Penfrel.

One of the major concerns of Hassan and the commercial group was the fact that within the next five years, a number of blockbuster antibiotics would lose their patent exclusivity around the world – they would become generic and cheap. Hassan and his team wondered how new antibiotics at higher price would compete under these circumstances. To Daniel, the explanation was obvious. New antibiotics had always competed well, even if they were not all billion dollar sellers, because they always offered some advantage – sometimes a small advantage – but usually something that was enough to drive sales. For example, the sulfa drugs had been around since the 1930s. The latest twist on those drugs was the combination drug, Bactrim which included a drug called trimethoprim in addition to the sulfa. Bactrim was active against sulfa-resistant bacteria. But Bactrim, because it contained a sulfa drug, was problematic because in rare cases it could cause fatal skin and liver reactions.

Erythromycin was marketed in the 1950s, but 30% of patients experienced gastrointestinal upset when taking it. Azithromycin and clarithromycin were drugs from the same chemical series as erythromycin that did not cause GI upset – they were both blockbusters with sales approaching $1 billion.

Over the years since the 1950s, many different cephalosporin antibiotics (related to penicillin) had come to the marketplace. Each one took their turn going generic, but not before its successor would come along. Each new cephalosporin had an advantage over the previous one. Sometimes they added additional bacteria to their spectrum of activity. Sometimes they decreased the number of doses per day that would be required. There were a few that were available in pill form as opposed to the normal intravenous form for these drugs. One could be given intramuscularly. Most achieved some measure of market success with a couple being clear blockbusters in spite of existing generic versions of older cephalosporins.

Daniel convinced Hassan to do a 25-year retrospective study of the effect of antibiotics like the sulfa drugs, erythromycin, cephalosporins and everything else going generic on the overall antibiotic market. When the results finally came in, they clearly showed that, as Daniel had predicted, there was absolutely no effect of generics on the dollar volume of the antibiotic market. As long as you could bring a new antibiotic to market that had some important if small advantage over the old ones, you could sell it at a premium price compared to the older generic antibiotics. Hassan was surprised and enthusiastic about his study. These data would definitely be shown during the infectious disease review in May.

The presentation would be based first on an overall market analysis both of antibacterial drugs and of antiviral drugs. Then, the medical needs for new therapies in these areas would be reviewed where antibacterials for drug resistant infection and the need for new, non-toxic therapy for Hepatitis C virus would be highlighted. Then, Penfrel's marketed antibiotic, peracillin would be reviewed emphasizing its market potential of over $1 billion which would be achieved soon in spite of all Penfrel's manufacturing problems. Teracil would be the product that would take the reins in the hospital after peracillin with teracil's albeit exaggerated projection of $1 billion in peak year sales as well. Earlier projects would include the new B-lactamase inhibitor program and a program targeting hepatitis C where Penfrel had engaged a collaborating biotech company called Viron. Daniel would have to explain Penfrel's failing effort to discover new antibiotics by exploiting the bacterial genome and to show the committee where the discovery effort would go in the future to find new antibiotics. While this was not the best topic for a review, Daniel felt that since Penfrel had invested in the project and since it was reaching its end without bearing fruit, some explanation and the presentation of some strategy for the future would be required.

The room was crowded. There were windows covering three walls. The sun was so bright that the room was uncomfortably warm, bringing out the sweat even more than Daniel's nerves had already done. It was difficult to see the slides Daniel and Hassan had brought with the sunlight streaming in in spite of drawn shades. Yosemite Sam and Penfrel's CEO, Hank Wallace, sat side by side. Next to them were the global head for US sales, the global head for Europe, the global head for Asia and rest of world and Mark Mann, the head of new products marketing. Then there were all of Sam's minions including Ib Hassad from manufacturing, Ralph George from drug safety, David Schubert from vaccines and a few folks from the infectious diseases therapeutic area including Larry Ryan and Richard Noland. Most sat around a large U-shaped table while those less significant others sat in surrounding chairs. Daniel tried to guess from the demeanor of key individuals like Sam and Mark Mann how the day would go. But everyone exuded an incredulous amicability that Daniel had never encountered before. In spite of the smiling faces and firm handshakes, there was no doubt in Daniel's mind that the days of infectious disease at Penfrel were numbered regardless of the outcome of this day's meeting.

Hassan and Daniel took turns presenting. Hassan led off with an overview of the anti-bacterial and antiviral markets. He covered in some detail how Penfrel's peracillin fit within the antibiotic marketplace and showed its sales projections. He then spent a fair amount of time presenting the study showing the lack of effect of generic intrusions into the antibiotic market. The only person who seemed to be surprised was Mark Mann, Hassan's boss. But the rest of the commercial heads were all nodding their heads. Daniel realized it was because Mark Mann, who had essentially no experience in actually selling an antibiotic, thought that the upcoming loss of blockbuster antibiotics to generics might be a huge problem. Apparently he was the only one in the room worried about that. After one or two questions, Mark realized this himself and was silent for the rest of the meeting.

Daniel covered the areas of medical need followed by Hassan on commercial potential for exploiting those areas of need and the kinds of drugs that would be required. Daniel picked up again to talk about how teracil would follow in the footsteps of peracillin by treating infections caused by resistant bacteria in the hospital. He outlined the plans for the development of teracil and reminded them of their successful battle with FDA suggesting that this could be repeated for other drugs in the future. Daniel couldn't have been more wrong on that point – but who knew? Hassan then went over the commercial projections for teracil.

Daniel then introduced the new program to find new B-lactamase inhibitors as a follow-on to peracillin. He highlighted the role of the chemists at Penfrel Japan and he showed how much progress they had made already. He reiterated their promise to deliver a candidate for clinical development within that year. Hassan picked up the thread by showing how, by dividing the market between a hospital-wide use drug like teracil and a new peracillin like drug more targeted at the intensive care unit and at a higher price, there would be room in the market for both. Daniel joked that it was a tough problem to have – having two promising antibiotic candidates one of which might come a few years after the other. Hassan noted that the Penfrel antibiotics pipeline would be one of the best in the industry.

During all this there were very few comments or questions. Daniel didn't know how to read this. It was possible that everyone in the room had, through their underlings on the infectious disease therapeutic area, already heard not only a preview of what was to be said, but had also provided input to the presentation.

There was a coffee break. During the break, most of the management team, Sam, Hank Wallace and their underlings hung out together. This seemed to be a rare opportunity for them to interact with the Hank. It made Daniel wonder how and from whom Hank got his information about what was going on in the company. Daniel had the thought that Hank's primary contacts within the company might be the CFO and the global marketing heads except in the case of problems. Yosemite Sam did come over to tell Daniel and Hassan that they were dong a great job. Daniel smiled and thanked him but thought about his future without Sam and Penfrel.

After the break, the antiviral strategy was next. Hassan presented a detailed analysis of the Hepatitis C market. He explained that in spite of the dramatic decrease in new cases of HCV, there was a very large reservoir of untreated patients who would, in the future, need therapy. Daniel then described the medical need for new HCV treatment emphasizing the toxicities of interferon therapy. Daniel outlined the Penfrel approach to discovering drugs with which to treat Hepatitis C virus. He had elected against bringing up the failure to include use of the chimp model as part of the Penfrel collaboration with Viron, but rather emphasized the progress that had been made in the collaboration and noted that their first drug was scheduled to start human trials later that same year. He then went on to talk about other viruses being explored by his team including Herpes zoster, the cause of chickenpox, but more importantly, the cause of shingles in adults. The medical need here was to find a way to prevent the severe chronic pain that could follow resolution of shingles since shingles can damage the nerve endings in cells that it attacks. The theory was that more effective antiviral therapy might alleviate or ameliorate this devastating complication of infection. This became moot later after the discovery that the chickenpox vaccine for children would prevent zoster in older adults. But at the time, new drugs for shingles were still needed.

Hassan got up to summarize the day painting an enthusiastic and exciting picture of the activities in the Penfrel infectious disease therapeutic area. He emphasized that peracillin would be a blockbuster and the teracil had that potential as well. He noted that generic intrusions in the antibiotic space had no effect on the market overall as long as products that could be differentiated could be introduced on the market. He then summarized the antiviral possibilities emphasizing the Penfrel-Viron collaboration and their first clinical candidate for that disease.

It was over. Daniel, Hassan, Larry, Richard and the others not on this special committee were asked to leave the room. Only Daniel and Hassan would be asked to wait while the committee deliberated. They promised feedback for them within the hour. Daniel and Hassan held their own post-mortem in a conference room next door. Both wanted nothing more than to head out for a drink. They both felt good about their presentation. On their way out to head home, many of the others who had been in the room for the presentation stopped by to congratulate them on a job well done. But no one knew how Hank Wallace, Yosemite Sam or any of the others were reacting. Daniel knew that this was going to be mainly between Sam and Hank and he would never put any trust in anything coming from Sam under any circumstances.

After only an hour or so, Daniel and Hassan were asked back into the room. The committee members were standing around sipping coffee and munching on cookies. Everyone sat down. Sam expressed his thanks and those of the rest of the committee. Hank Wallace said his first words of the day to express his thanks as well. Sam said that the committee had deliberated and that they were in overall agreement with the strategy enunciated by the infectious disease therapeutic area. They were especially enthusiastic about teracil and the HCV collaboration. Hank Wallace got up to shake Daniel's hand before leaving. "Thanks for your hard work," he said. "That was a great presentation." And that was that – almost. As everyone was heading back to work, or in Daniel's case, to the nearest bar, Sam grabbed Daniel's arm. "Great work!" he said. "Infectious disease is here to stay at Penfrel. Congratulations." Daniel headed to the bar down the street, ordered a beer and called all the key members of the therapeutic area with the news. Everyone except Daniel was ecstatic. Sadness and frustration were his only emotions following the meeting.

Within six weeks, Daniel would be working at another company in Boston. Within six months, Penfrel would pull out of infectious disease and fire most of the people who had worked for Daniel. Daniel always wondered what would have happened if he had stayed. He felt like this massive loss of jobs was somehow his fault. But, on the other hand, he knew that this would have come whether he was there or not. The decision had been made back in October the year before they ever had their review with Hank Wallace.

In retrospect, when Daniel looked back at the history of the pharmaceutical industry, he was awestruck at the number of jobs that had disappeared over the years. And this was not at all isolated to jobs in antibiotics. Penfrel itself was the product of over twenty acquisitions or mergers over the preceding 30 years. Each one of those was associated with job losses to provide "synergy" for Wall Street. The same was true for all the large pharmaceutical companies then extant. Then, when Daniel thought about how many large companies had abandoned antibiotics research per se over the last few years, he realized that antibiotics had been hit with the double whammy. Mergers and acquisitions had led to fewer companies. Then the loss of companies pursuing antibiotic research occurred. These two trends resulted in the disappearance of several thousand jobs for experts in antibiotic discovery and development. Who would do this research in the future? Where would the new antibiotics come from? It was clear to Daniel that they were not going to come from large pharma.

Penfrel's teracil, though, would make it to the market. It would be a much more modest success than what was predicted by the Penfrel team. It was even more modest that the predictions of Daniel and Roger Butler. But it did offer another alternative for therapy of very highly resistant infections and helped physicians and patients hold out while they waited for the next new antibiotic that would be active against these infections. The increase in sales of teracil paralleled the rise of these resistant infections. The next antibiotic with activity against these infections wouldn't arrive until ten years after the launch of teracil.

Chapter 14

Looking Glass

While at Penfrel, Daniel lost one of his best friends to leukemia. Bernard and his wife Suzie lived in Paris and they had become good friends during Daniel's sabbatical there in the late 1980s. They were exactly the same age with birthdates in the same month and year. Daniel never even heard about Bernard's illness. He called on one of his routine trips to Paris for Penfrel only to hear the news from his widow. Daniel was inconsolable.

Before Daniel left Penfrel, George Finkel fell ill. He had been trying to retire from Penfrel for several years – almost since he had been moved aside during the reorganization where Penfrel replaced him with Brian McKinley. George wanted to spend more quality time with his family. He had a son and daughter-in-law and two grandchildren living in Delhi where his son was working for USAID. He had two other sons and two other grandchildren scattered around the US. George had been in his awkward advisory position for three years when he began to experience stomach pains that would not go away. He thought at first that it was an ulcer or heartburn. When he finally went to see a physician a few weeks later when the usual treatments were not working, he was found to have inoperable pancreatic cancer. He was dead six weeks later. Daniel and Sally went to the funeral. Daniel resolved then that he was not going to end up like George if he could help it. When it came time for Daniel to retire – he would.

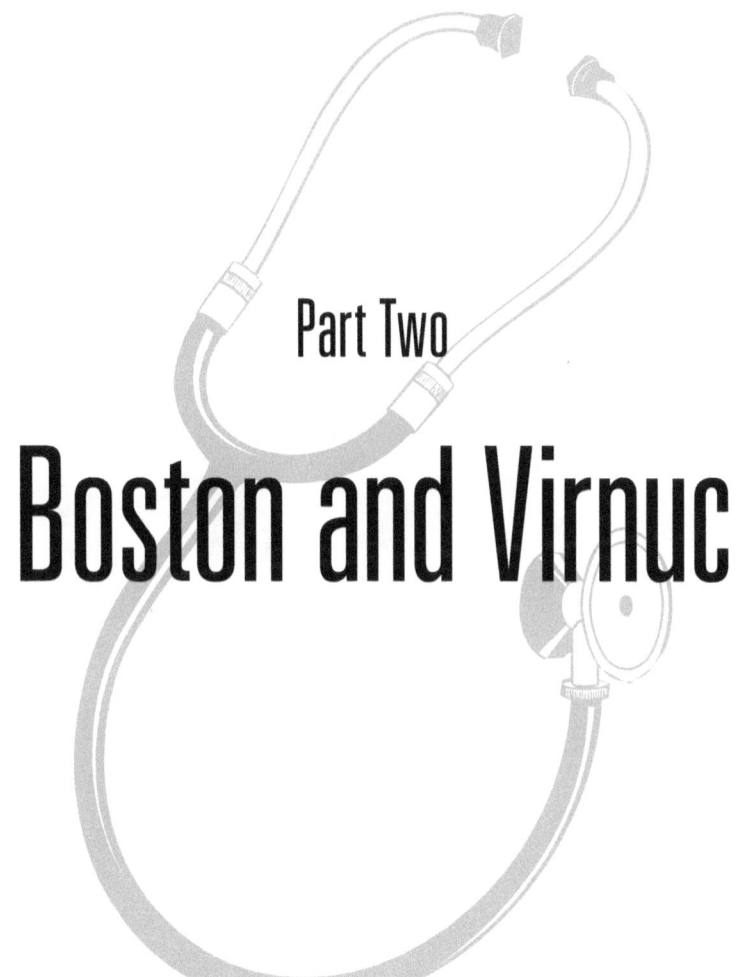

Part Two

Boston and Virnuc

Chapter 1

Boston

After the therapeutic area review at Penfrel, with the handshakes from Hank Wallace and Yosemite Sam, Daniel once again began thinking about his future. The stress of the corporate struggles in a large organization like Penfrel was depressing for Daniel. This was not helped by the travel to Collegeville with a routine of ten days on the road per month. Finally, Daniel didn't know how he could advance within Penfrel. He wanted to stay involved in infectious disease. That meant that larger opportunities like heading all of research and development or even getting more involved in clinical development, as career paths, were unlikely to come his way. This was especially true with Yosemite Sam firmly in control of research and development at Penfrel. These questions were boiling up through Daniel's thoughts in May of 2002. Daniel had confided in Bos Killington ever since the phone call from Sam Stern the previous October.

Daniel received yet another call from another headhunter. In the six years Daniel was at Penfrel, headhunter calls were routine. He received several every month, mostly about positions that did not interest him at all. He was relatively happy at Penfrel. But he was always polite and tried to be helpful. You never know. Daniel did in fact look at two job opportunities during his years at Penfrel prior to 2002. One was at Abbot. They were looking for a new head for their anti-infectives discovery group. He interviewed for the job, but a few weeks later they decided to abandon the entire field of antibiotic research and development. Easy come easy go.

Another job possibility was at Novartis. This came up shortly after the merger between Ciba-Geigy and Sandoz that resulted in the formation of Novartis. Daniel again was looking at a head of anti-infective discovery there. They wanted to establish a new discovery group in California at that time. Again, shortly after his interview, their plans changed and they delayed everything to establish what was to become their Biomedical Research Institute in Cambridge, MA. They eventually established an anti-infective research group there, but that was years after Daniel's interview with them.

But in May of 2002, the headhunter was asking Daniel about his interest in a small biotech company focusing on antiviral drug discovery in the Boston area. The job was to be their chief scientific officer, which meant that he would be responsible for drug safety and pharmacology, manufacturing, and all their drug discovery research. The company, Virnuc, had gotten Daniel's name from Bos Killingworth. This got Daniel's attention immediately since he thought that Bos would never steer him wrong. Virnuc had research centers in Lexington, Massachusetts just outside of Boston, in Italy and in France.

Aside from Bos' apparent blessing, there were a number of attractive features to the job. First, Daniel would finally be able to advance his career by taking on new and very important responsibilities. He would have to learn a good deal more about viruses and antiviral drugs but, on the down side, he would be out of the antibiotics business. He would essentially control everything within research and development at Virnuc except clinical development and regulatory affairs. He would report directly to the CEO whose office would be one or two down the hall from Daniel's. He quickly agreed to meet the CEO for an interview in New York.

Before heading to New York for his interview, Daniel undertook some research on Virnuc and its CEO, Roman. First he called Bos. "Well,"

Bos said, "Roman is not the easiest person in the work to work for. He has a big ego. But I think you can deal with it. Your problem will be the loss of antibiotic research."

Daniel's internet research on Roman indicated that he was very well known in the antiviral arena. His name was Roman. He was born somewhere in Russia – Daniel never knew exactly where. Roman and his parents immigrated to France during his early childhood. His parents then divorced and his mother married an Italian man. Roman spoke Russian, several other Slavic languages, French and Italian fluently. But he preferred English. On the rare occasions when Roman and Daniel spoke French, Roman would throw in lots of English words to substitute for the French words he had forgotten. Roman came to the United States after completing his doctoral work and post-doctoral training in France. Eventually, he ended up as a professor at Emory University and had been a colleague of Bos' for a number of years.

Roman seemed to be a well-respected and talented scientist who had been consulted by most of the major drug companies before he left the university to establish his own company just a few years ago. Daniel called a few other friends in the antiviral area. They told him that, in fact, Roman was an outstanding scientist with innovative ideas. They also warned him, like Bos, that Roman had an outsized ego. But the size of Roman's ego did not scare Daniel. After six years at Penfrel, Daniel was used to working with big egos. Daniel was wrong not to be worried.

Daniel and Sally spent many evenings talking about this. Sally was thrilled with the idea of living in the Boston area since she had a brother and sister living there to say nothing of Sally and Daniel's daughters both in Cambridge. But she was worried about Daniel's proposed new job. "This guy Roman doesn't sound like he'll be a great boss. And are you ready to switch over to virology? What happened to antibiotics?" Sally clearly had Daniel's number on this.

"I'll see what Roman is like after I meet him – but its not like I haven't been dealing with this sort of boss for the last few years anyway.

I might like learning virology and at the same time I'll learn much more about lots of different areas like pharmacology, toxicology and even manufacturing. This could be really cool."

Sally said she was game if Daniel felt like this was the right thing to do.

Daniel drove into New York and met Roman for dinner at Le Cirque. The evening did not start well since Daniel was sitting in their plush lounge with a glass of wine waiting for Roman while he was already at the table waiting for Daniel to show up. This went on for about 30 minutes before Daniel asked the Maître d'hôtel to check again.

Roman was tall, thin, just over 6 feet, had dark, straight hair, a prominent nose and full lips. Daniel once asked him how he kept so thin. "I only eat one meal a day." But that obviously did not count the chocolates he had constantly available in his office nor the multiple cups of espresso, from a very expensive machine also in his office, which he consumed with a great deal of sugar.

Roman had a slight accent when he spoke English. Later, Daniel would hear a slight accent when he spoke French as well. But Daniel couldn't place the accent at all. It didn't sound Russian or Slavic and didn't sound Italian. It might have been like a southern French accent, but Daniel was never sure what it was.

Daniel asked why Roman was interested in him. "I am an antibiotics guy." Although antiviral discovery was part of his job at Penfrel, and although they had completed a multimillion-dollar deal with another antiviral company, Viron, under his leadership, Daniel did not consider himself an antiviral expert. Roman explained that his company, Virnuc, was about to go public. It employed about fifty people worldwide compared to the 50,000 or so at Penfrel. He wanted to use some of the funding from the public markets to expand into antibiotics. To him, Daniel's expertise was critical. Of course, this was music to Daniel's ears after having just spent six months of his life working days, nights and weekends to prevent Penfrel from canning their entire antibiotics discovery effort.

Roman also assured Daniel that he would have Roman's full support while he went through the learning curve on virology that he knew Daniel would have to climb. Their conversation went on into the night with both plying each other with questions. Roman waxed poetic about the future of his small company. He talked about their drug for Hepatitis B that was already in late stage trials. And he discussed a new drug for Hepatitis C that had just been tested in chimps and that worked without toxicity. He drew graphs and tables on drink napkins. Roman did all this without the usual confidentiality agreements companies like to have in place before discussing their internal results. Maybe Daniel should have been a little worried about this – but he wasn't. Roman's excitement was contagious. At the end of the evening, it was approaching one a.m. or so, he outlined a job offer on another drink napkin. He said that Virnuc was in a big hurry because their initial public offering was imminent. If Daniel wanted to be able to cash in on stock options, he would have to decide soon.

On getting back home, Daniel had a lot to think about. In discussing it all with Sally, they were both excited about the possibility of moving up to the Boston area. Both of their daughters as well as one of Sally's brothers and her sister all lived in the Boston area. And they were attracted to the idea of living in a more suburban environment again as opposed to rural, and somewhat isolated, New Jersey countryside. Sally's worries about the job did not abate, though. She was not so reassured by Daniel's interview with Roman.

The offer arrived within a week. It was substantially what Roman had outlined on the napkin at Le Cirque. But Virnuc wanted Daniel to make a decision within two weeks. Daniel engaged a human resources lawyer for $500 per hour on the advice of Bos. The lawyer made a number of suggestions for modifications to the offer, but Roman was having none of it. It was, mostly, take it or leave it. But, when Daniel thought about it, the offer was a good one. He would not get a rise in salary – but he was already making plenty of money. Even considering the increase in cost of living in Boston compared to New Jersey, Daniel and Sally wouldn't suffer. He would get those potentially valuable stock options, and, as Roman said in Le Cirque – it's the stock options that count. Daniel wasn't so sure. For the stock options to count, the company would have to go public, be bought by another company or both and none of those things had yet happened.

But many of Daniel's career goals would be met. He would have very expanded responsibilities including manufacturing, pharmacology and toxicology, and all of discovery research including chemistry and scale up chemistry research. Plus the sites in France and Italy would be under his control. Or at least that was what the offer said.

This offer in combination with Daniel's fears about what Yosemite Sam was going to be doing with infectious disease at Penfrel was too much for Daniel to refuse. Of course, this meant that Penfrel would have only very short notice. It also meant that Daniel and Sally would have to say a very quick good-bye to all their new friends and colleagues with whom they had established strong and sometimes close relationships over the previous six years.

The Penfrel scientists held a big luncheon get together at the local bar. Beer flowed. People were sad to see Daniel go, but they were glad to be able to keep on working on antibiotics at Penfrel. Little did they know that with his departure, in spite of Daniel's handshake with the CEO, Penfrel would fire almost all of them within the next six months.

Daniel arrived at Virnuc in mid-June. He stayed at his brother-in-law's apartment while Sally handled the sale of their house in New Jersey and arrangements for the move. Daniel was charged with looking for a house in the Boston area. Mark, Daniel's brother-in-law, rented a duplex apartment that was one half of a small house owned by Mark and Sally's sister, Daniel's sister-in-law, Maureen. Maureen occupied the other apartment next door. Maureen's place was crawling with four cats. Daniel was very allergic to cats and was unable to stay in Maureen's apartment for more than an hour or so before clogging up completely. So he spent most of his time while not at work in Mark's place. Daniel and Sally's daughters both had small apartments. The older daughter also had two young children and a husband. There was no way Daniel could stay with them but he saw his daughters and grandchildren frequently.

Daniel and Sally would commute to New Jersey or Boston on the weekends. Sally came up for longer stays as Daniel identified possible houses to buy. The problem was that there was nothing they could really do, other than be aware of the market, until their house in New Jersey was sold. In fact, the market in Boston at the time was such that houses did not stay on the market for longer than a few days and prices were frequently bid up above the asking price before they were sold. Because of buyers whose loans could not get approved and other problems, they were unable to sell their home in New Jersey before October.

They found a house they liked in a Boston suburb at more than twice the price of their New Jersey home. They were able to see the house on its first day on the market and offered just under the asking price. The sellers accepted their offer and by some miracle, the sellers seemed to like Daniel and Sally, they refused other higher bids that came in later. Daniel and Sally would finally move in in November, five months after Daniel had arrived at Mark's place in Boston.

Their first houseguests outside of family would be Roman, his wife and daughter who came over for Thanksgiving. Daniel and Sally's family was there including both girls, two grandkids ages 3 and 18 months, Mark and his girlfriend and Maureen. Roman brought a case of very expensive and very good French wine. Things were starting to come together for Daniel in Boston. Sally was ecstatic.

The house they bought was large and was originally built in the 1930s. It had recently undergone a complete makeover in the Arts and Crafts style of the turn of the last century. Wood was everywhere including floors and walls. The ceilings on the ground floor were high and there were large windows everywhere. So in spite of all the dark wood, the house was light. The major problem with the house was that it was located on a busy corner with a traffic light. Road noise, in spite of the double-paned windows, was constant. In the warm weather, Sally and Daniel used the air conditioning rather than opening the windows to avoid the road noise. But Sally had her garden and was content. Daniel initially worried about the size of their mortgage – but they never seemed to suffer financially.

Roman invited Bos up to Boston ostensibly for a day of meetings since he was a Virnuc consultant. But his real motivation was a celebratory dinner with Daniel. Roman arranged a dinner at an exclusive Boston restaurant where they specialized in French wines found at auction. They started with Krug champagne and foie gras. They moved on to a bottle of 1976 Chateau Latour. Desert involved an old, rare Chateau Yquem. They were at the restaurant from 8 pm until after midnight. The bill was close to $3000. Roman paid with his personal credit card – he didn't bill Virnuc. Daniel was impressed and glowed with a warm feeling beyond the alcohol.

After arriving at Virnuc, Daniel once again was treated to project reviews. This time, though, there were only three projects and the team was already very focused. The first was work around the hepatitis B drug that was already in late stage trials. The Virnuc team was trying to understand whether the drug, teldine, would select for resistance. Hepatitis B, like Hepatitis C, was a disappearing disease. There was a very effective and safe vaccine that was being used worldwide to inoculate children in areas of the world where the infection was already widespread, and to inoculate children and adults in other countries where the disease was confined to those at high risk like healthcare workers. But for those already infected, the drugs available to control the infection rapidly selected for resistant virus and became either ineffective entirely or at least partially where the patients were subject to violent viral breakthrough episodes which could be life threatening. Since teldine worked slightly differently than the drugs already marketed, the scientists had to determine whether it too could select for resistant virus and if so how frequently. This work was still ongoing when Daniel arrived, but it was clear to Daniel and to the Virnuc scientists that teldine would not be immune to resistance. This was a worry, since there were two other hepatitis B drugs in development from large pharma competitors like Bristol-Myers Squibb that promised to be almost resistance-proof. Teldine would be safer than these competitors, but whether it too would be resistance-proof seemed highly doubtful.

The second project involved a so-called backup to teldine. The current idea was to combine this drug, called cydine, with teldine for those unfortunate individuals who already had virus resistant to both teldine and other drugs, assuming that teldine resistance would, in fact, be problematic. The problem, Daniel learned, was that cydine was not nearly as safe as teldine, and could only be imagined for patients in somewhat desperate straights willing to risk a certain amount of toxicity. And, if one of the resistance-proof competitor drugs made it all the way to the market, there would probably be no need for cydine.

But since neither Roman nor anyone else knew whether these other drugs would make it, Roman wanted to push cydine development along and he wanted to start a trial looking at the combination of cydine and teldine for sick patients with resistant virus infection. This was a very tall order since it would involve a clinical trial combining two drugs neither of which had yet been approved. There was not a clear pathway for the regulatory agencies to allow you to do this. One of Daniel's first tasks would be to develop such a plan and get agreement for the plan with the FDA. Daniel's reaction was, "good luck with that one!"

The third project, and perhaps the most important one at Virnuc at the time, was the drug against hepatitis C virus that Roman had discussed with Daniel during the interview at Le Cirque. This drug targeted the HCV's ability to reproduce in cells by hitting the key enzyme responsible for duplicating the HCV nucleic acid, the polymerase. It was called CI-999 –what this stood for, Daniel never knew. CI-999 worked on the HCV replicon in cells at a reasonable potency similar to the potency achieved by the earliest drugs against the virus causing

AIDS, HIV. And, as Roman had drawn on a cocktail napkin, the drug had been shown to work in chimpanzees. The experiment involved five chimps chronically infected with HCV. One chimp received no therapy, two chimps received a lower dose of CI-999 and two chimps received a higher dose. There was a dose response on average, but there was a fair amount of scatter in the data. It was crystal clear and beyond a doubt, though, that CI-999 had an antiviral effect in the chimps. Daniel was convinced that it would work in people too if they could achieve the dose levels they were able to achieve in the chimps.

One of the great things about HCV was, that unlike hepatitis B and the AIDS virus, HCV had nowhere to hide. It could not lie in a dormant state and avoid the effects of therapy. HCV drugs would not control disease if they worked, they would cure the disease. The problem was that HCV mutated so quickly and so often that resistance was likely to be present at the beginning of therapy, so some sort of combination with the existing drugs, toxic as they were, was going to be necessary just to avoid this resistance. They were unlikely to get resistant to everything all at once.

When Daniel arrived at Virnuc, these data were all in. Virnuc was awaiting two pieces of a document they were preparing to send to the FDA – the Investigational New Drug application or IND. The top priority for Daniel as outlined by Roman, was to get this document submitted before November. The two pieces that were missing were the toxicology section and the clinical development plan. Daniel was responsible for the toxicology section.

The toxicology tests were running for two weeks in rats and monkeys and the data was not expected until sometime in late October. Daniel didn't understand why, if the tests were started in June and only went on for two weeks, they would have to wait for another four months to see the data. The reason is that the dosing of the animals is the shortest and easiest part of the testing. The animals had to be sacrificed, autopsies performed and organs collected to be examined under the microscope. Blood had been taken during the two week dosing and at the time of sacrifice to estimate drug levels in the blood and to look for abnormalities in the blood counts and blood chemistry. All these tests were done in batch when the contractor could get them done. Then the data had to be reviewed such that it was "quality assured." This is a sort of guarantee that there are no mistakes in the actual recording of the data in the report supplied by the contractor. Daniel would not submit a document to the FDA without being sure that all the data was guaranteed not to change – the alternative could be embarrassing and costly. Then, the report had to be written, a summary had to be generated and the report would ultimately be finalized. But you don't need a final report for the FDA. A summary with QA data was all you needed. Virnuc virtually never bothered to finalize their toxicology reports until they were ready to submit a marketing application to the agency after all the studies had been completed.

The timeline for the toxicology report meant that they would only have about two weeks to complete writing the IND before having to make sufficient copies and get them bound and sent overnight to the FDA. The clinical development plan, Daniel thought, could be written anytime. It did not really have to wait for the toxicology report. But Mary O'Brien, the clinician responsible for writing the plan disagreed. She argued, correctly, that she could not estimate her starting dose for the trials in volunteers until she knew the dose of the drug that gave no toxic effect in the animals. Daniel thought that the plan for dosing volunteers was so cookie-cutter that it could be written ahead of time leaving the dose levels to be added later. You would first give volunteers single doses of drug, look for side effects, and if there were no important safety signals, you would move on to the next dose. Sometimes you had to obtain blood samples to look for drug levels before going to the next dose depending on the drug. In this case, that would not be necessary – the drug levels could be examined in batch at the end of the trial. Once the single dose was completed, you could take that safe dose and go ahead and administer it for the 14 days of repeat dosing. CI-999 was in pill form, so this was all fairly straightforward. Daniel figured you could complete the entire trial in just a few months by staggering the single and multiple dose levels. But that was not the way Mary did things. First you completed all the single dose studies and only then could you move on to the multiple dose studies in her paradigm. But there was no reason for her paradigm either in regulations or anything else. Daniel thought Mary was just asserting her control over the process for the sake of her own ego – but there was nothing he could do. He did try and reason with her and her boss, the chief medical officer for Virnuc, Matt Boden.

Matt seemed to be an easygoing guy during Daniel's early conversations with him. Matt was probably the best and most experienced hepatitis drug developer in the industry. But he would later learn that Matt was unable to delegate to anybody and had to do virtually everything himself. Even when it became clear that he could not handle the number of late stage trials that Virnuc was undertaking and even after reluctantly hiring two other physicians to help him and Mary, Matt was still doing most of the writing and making most of the decisions. Daniel could not imagine actually working for him. But when Daniel first arrived, Matt was all smiles and seemed relaxed. When it came to Mary, though, Matt knew his limits. He had tried this in the past and decided that retreat was a lot better than constant confrontation. Daniel adopted the same approach.

It was clear to Daniel, from the outset, that on everything involved in drug discovery and development, Roman had surrounded himself with real, usually well-known, experts in their fields. Those people that reported to Roman, including Daniel, Matt, the chief financial officer, the lead counsel, were all highly respected experienced and senior experts in their own areas. It was also clear from the outset that Roman made the decisions, frequently ignoring the advice of all the smart people with whom he had surrounded himself. Matt and Daniel questioned him on several occasions as to the wisdom of trying to develop a combination of two experimental drugs, in this case cydine and teldine, before either one had been approved. There were many reasons why this was a bad idea – the main one being that one drug in the combination could put the other drug at risk when some problem turned up in the combination and you didn't know which drug was responsible. Teldine was well on its way to getting to market. Cydine might kill teldine's chances. But Roman imperiously overruled everyone and insisted on going forward with this plan. The executive committee meetings basically became meetings where Roman would tell everyone else what they would do.

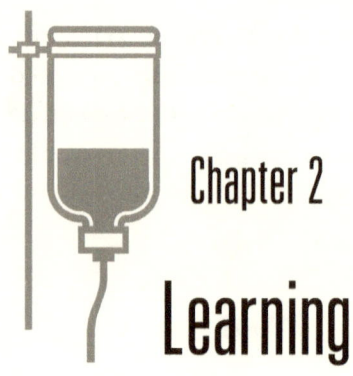

Chapter 2

Learning

On arrival at Virnuc, Daniel would once again be thrown into a job where he would have to direct activities that he knew little about. To do this, he would have to learn a great deal since, at the end of the day, he would be held responsible. His greatest challenge in this regard was manufacturing. It is true that while at Penfrel, he had to learn enough about manufacturing to deal with those carrying out those activities. But that level of understanding is much different than that required to actually direct these activities. In a drug company, you would think that Daniel would have to become an accomplished chemist or chemical engineer in order to direct this kind of manufacturing. But what he really had to do was hire good people who were good chemists and chemical engineers and who could explain what they were doing such that Daniel could understand it. They could then be trusted to direct these activities such that Virnuc goals and timelines were met. Daniel's job then became one of garnering management support for his head of manufacturing.

When Daniel arrived, Brad Baron was a chemical engineer who was already in place directing Virnuc's manufacturing efforts. Brad looked like a linebacker for Green Bay. He had crooked or missing teeth, was bald with a shaved head and was about 250 lbs but under six feet tall. At the time, Virnuc's needs were not large since only one drug was actually in clinical trials. But Brad had not only to direct manufacture of the drug powder (API) but also manufacture of pills or capsules made from drug powder. For Virnuc's early trials, they just manufactured capsules that were filled with drug powder or, for the placebo, that were filled with sugar. Brad was clear, decisive, and extremely competent. Brad was also a good teacher and could explain what he was doing in detail such that Daniel, in turn, could explain things to Roman and to the Virnuc Board if necessary. Brad had been hoping that he would have been promoted to a higher position within Virnuc, but Daniel's arrival stifled that possibility. But Brad and Daniel got along well in spite of the fact that occasionally, Roman would go directly to Brad with questions instead of passing them through Daniel. Daniel made sure to reward Brad as things progressed.

The major problem with the manufacturing activities was not Brad or Virnuc's manufacturing contractors – at least early on. It was Mary and the group within clinical operations. The clinops group under Mary was supposed to provide Brad with a plan specifying how many capsules would be needed by which clinical trial sites by which dates. In this way, Brad would be able to time his manufacturing and the entire supply chain. His job was to get the correct number of capsules to each of several regional distribution centers, which, in turn, were responsible for supplying trial sites within their region. When Daniel arrived, this should not have been difficult since there were only a few trial sites and most were in Hong Kong and Taiwan where the hepatitis B trials of teldine were ongoing. But Mary and her group could never get the information to Brad on time and Brad was tearing out what little hair he had left. Daniel was a little surprised since Brad did not look like someone you wanted to get angry. But Mary, apparently, was afraid of no one.

Daniel could see what was coming down the pike as Virnuc planned more and more trials in more and more sites around the world. He quickly hired two other members of the manufacturing team. The first was Morris Townsend, a tall, lanky, gray-haired British gentleman with a gray mustache. Morris was an expert at making drug powder into pills. Daniel knew that drug powder in a capsule could not continue beyond early phase clinical trials and that Virnuc would need pills.

Pills were not so simple to make. The powder had to be mixed with other ingredients to allow it to flow smoothly in the machines used to form and compress it into pills. At the same time, these other ingredients could not interfere or interact chemically with the drug. All this required time to experiment with different mixtures to find just the right one for any given drug powder. Things were actually even more complicated since the physical form of the drug powder- which crystal form, how many crystal forms, etc. - were all important considerations. Morris was an expert in all of this and Daniel was grateful to have him on board.

Predictably, as cydine and later CI-999 entered clinical trials and advanced to more complex trials, the supply chain issue became critical. Although Morris was an expert in how to make pills from powder, running a supply chain, especially one where he would have to confront Mary on a regular basis, was not his forte. Daniel quickly figured this out and hired Laura Coleman. Laura had been working at a company in Framingham, Massachusetts – not far from Boston. Her job involved managing supply chains for manufacture of components for hospital products. When she interviewed, Laura impressed Daniel as someone who could take on Mary and the clinops group and succeed. She also had the critical experience of supply chain management. He hired her after her first interview. Laura lived up to her promise. She worked with Mary and was much more successful that Daniel, Brad or Morris had been. When things did not work well with Mary or when Mary was tied up or otherwise distracted (supply chain was not her only activity), Laura found others in Mary's department who were willing to help. Laura constructed a supply chain database that Virnuc used long after both she and Daniel had departed.

In his wildest dreams as a clinician scientist at the University in Cleveland, he had never dreamed that he would be worrying about getting drug powder from manufacturer A to B and pills from B to distribution centers C through K. But if you want to run clinical trials, you can't do it without drug.

Daniel learned about drug stability. He learned about transport. And he learned that no matter how well you do things, how carefully you monitor your contractors, shit still happens. In the middle of the late stage teldine trials, Morris came into Daniel's office one afternoon to announce that their contractor for the manufacture of teldine tablets had lost one entire batch of pills. Forty kilograms of drug powder at $10,000 or so per kilogram was down the drain. The batch came out cracked and unusable. An investigation into what had happened showed that the mixture had been heated somewhere along the line. This heating had altered the crystal form of the drug and this resulted in cracked tablets. The contractor identified a short circuit in the processing line and fixed the problem – but everyone was furious – and the targets of their anger were Daniel and Morris – not the contractor. Roman called Daniel into his office and berated him for half an hour. Roman later apologized. But with everything else going on between Daniel and Roman at the time, this was not helpful. Daniel would learn later that if that was to be his biggest manufacturing problem, he could consider himself very lucky indeed. It was. He did.

After Daniel left Virnuc, his experience in manufacturing there would be a major asset in his ability to advise other companies – especially small companies who tended not to think too much about manufacturing early on.

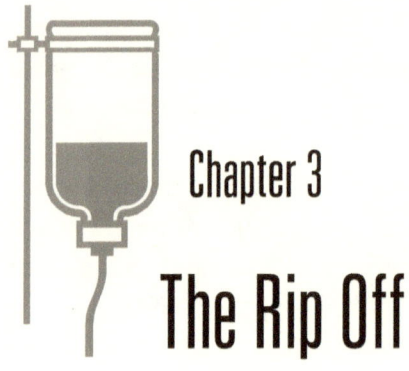

Chapter 3

The Rip Off

Within a few weeks of Daniel's arrival at Virnuc, towards late June, Roman arranged to take him to the research sites in Italy and France. He wanted to personally introduce Daniel to the groups there. As Daniel was preparing for this trip, he delved into the organizational structure of Virnuc especially as it concerned these European sites. The Virnuc international research sites were organizationally innovative, fascinating and dysfunctional. When Roman was a post-doctoral scientist working in a laboratory in Paris, France, he was joined by several PhD scientists from Hoffman-LaRoche. They were spending various periods of time working in the laboratory of Roman's post-doctoral advisor learning new techniques of working with viruses. Roman was inspired by this experience. When he founded Virnuc, Roman established two laboratories in Europe. One was located in Paris under the leadership of Frederic Laurent, one of the world's most highly respected chemists working on antiviral drugs. Roman hired about ten chemists in France and placed them under the tutelage of Frederic in Frederic's laboratory at the University of Paris. The other was at the University of Palermo in Italy. The universities provided all the ancillary support including library facilities, waste disposal, chemical storage, etc. In return, they would have limited rights to drugs discovered at the Virnuc facility within the University labs. Roman and Virnuc also kept a right of first refusal on any invention that might come out of the laboratories at the two universities. This meant that the university was obligated to negotiate with Virnuc for any invention that might be produced by the academic laboratories specified in the agreement.

In fact, two of the products being developed by Virnuc, teldine and CI-999 came out of the University of Palermo. The laboratories were each led by a professor. Virnuc paid the salaries of the Virnuc chemists, but they did not really have anyone from Virnuc as their supervisor. All the supervision including performance evaluations was left to the Virnuc consultants at the two universities. With the arrival of Daniel, the Virnuc employees would all become part of Daniel's overall group and would somehow report into him. But exactly how that would work awaited Daniel's trip and discussions between Daniel, the consulting professors and Roman.

Roman had a close friend in Italy who had an ongoing screening program looking for antivirals, anti-bacterial and anti-fungals. His name was Rico Allegri. Rico was not nearly as well known as Roman, Frederic, or for that matter, as Daniel himself. But, his laboratory had discovered CI-999 and Rico's name was on both the CI-999 and teldine patents. Daniel would learn much later that Rico's patent rights had never been signed over to Virnuc nor had they been assigned to the University of Palermo, on the island of Sicily in Italy where he worked. Virnuc had a parallel arrangement with the University of Palermo as they did in Paris. In return for the use of common university facilities, the University would get limited rights to discoveries made in Palermo. Virnuc had about the same number of employees working in Palermo as in Paris including a very able secretary, Lillia, who would become Daniel's girl Friday in Italy. Like Frederic, Rico was a consultant to Virnuc. And, as for Paris, the Virnuc employees in Italy would be moved into Daniel's group and under Daniel's supervision.

Before his trip, one of Daniel's senior employees, Diane, the project manager for CI-999 at the time, requested a meeting with him to discuss his upcoming trip.

Diane was clearly nervous about this particular meeting with Daniel. "I wanted to give you a heads-up about the situation with our scientists in Italy."

"Oh?"

"Yeah. Have you heard anything yet about the Italian research site?"

"Only what Roman told me. What's on your mind?"

Diane sighed. "This is hard to discuss. Rico is a lech. You'll see that virtually all the scientists working for him are young women. And that's not all."

"What does that mean?"

"The last time I was there, I walked into Rico's office to get some papers I left there before I went into a meeting with the screening scientists. I found Rico with his pants down around his ankles on the floor with one of his post-doctoral fellows who had her skirt up to her chest. She is maybe twenty-five years old."

Daniel's eyebrows went up. "Was this consensual?"

"It sure looked that way to me. But its not exactly appropriate behavior, is it?"

Daniel thought briefly before answering that one. "Of course not."

"When I mentioned this to Lillia, she told me that their affair was common knowledge and that Rico made no attempt to hide anything. The postdoc, Dorothea, was his frequent date for university and company functions."

"What happened after you peaked into his office?" Daniel asked. "Did they see you?"

"I don't know if they saw me. But everyone just acted as if nothing had happened."

Daniel reached for the pile of papers and folders that completely obscured his desktop. "Who is Dorothea? Is she a Virnuc employee? I don't see her on the list." Daniel said scanning through the file on the University of Palermo research site he had compiled in preparation for his trip.

"No. As far as I know she has some sort of grant from Italy for her post-doc in Rico's lab."

"Have you discussed this with Roman?"

"Are you kidding? They're best friends and have been forever."

Daniel knew that affairs between professors and students occurred. He disapproved, but if both were consenting adults and the couple was open about everything, what could or should he do?

"OK, Diane. Thanks for the heads-up. I'll keep my eyes and ears open."

Diane looked at Daniel. He had a feeling she was rolling her eyes as she turned away and left his office. But Daniel was worried about this.

Italy was Daniel's first stop on this trip. They left Boston on Alitalia, flying through Milan and arrived in Palermo on a Sunday morning. Roman spent the overnight flight working. Daniel tried to sleep. Roman wanted to spend the day sailing on Rico's boat. Daniel resented having to fly over on a Saturday night and miss out on a Sunday with his family. But this was the job, and he was still excited about his new responsibilities. Rico met them at the airport. Roman had checked luggage and it did not arrive in Palermo – apparently a typical Alitalia problem. Not only that, but Alitalia is very strict on the size of carry-on luggage – so Roman routinely checked a suitcase and Alitalia routinely lost it. It usually showed up a day or two later. Being a Sunday morning, nothing was open. Roman would have to do without a change of clothes and toiletries for a day.

Rico drove them up the coast through gorgeous countryside with dramatic views of the sea that reminded Daniel of the California coast. Here and there, there were red tiled roofs and small farms dotting the hills approaching the sea. Daniel was in the back seat of Rico's sports Mercedes and Rico was driving as if he were on a race course on curving roads that climbed and dipped constantly. In spite of the gorgeous scenery, Daniel was obligated to keep his eyes shut after the first 15 minutes or so to avoid getting carsick.

An hour later, they arrived at the marina where Rico kept his boat. It was a thirty-foot sailboat with a motor and two masts. The seas were moderate and Daniel was feeling the jet lag. But he managed to stay awake and avoid being ill for the day. They sailed for a couple of hours to a deserted island where they anchored and swam in the warm

Mediterranean. Rico brought a supply of sandwiches, beer and wine for the day. Daniel was thoroughly enjoying his day with Rico and Roman. Roman was constantly on his cell phone. Virnuc was discussing partnerships with a number of large pharmaceutical companies and Roman, of course, was directing the discussions personally. He was getting calls from lawyers, Virnuc's business development person and from the potential partners all trying to position themselves with Roman. But Roman knew what he wanted and kept his red lines red.

But this gave Daniel a chance to have a good discussion with Rico – not about Virnuc or Rico's lab, but just about living in Palermo, Italy, about families, food, wine, and Rico's future plans. Daniel guessed that Rico was about 65 years old already. Rico was not tall, was graying with medium length hair and had weathered skin. He was of medium build, had blue eyes and was constantly putting on and taking off reading glasses with thick brown frames. In most European universities, Rico would be at the age of mandatory retirement. In Italy, apparently, there was a way for Rico to keep working for a number of years after age 65 before retiring. And Rico had no plans to retire. He said that even though he enjoyed his family, he was divorced and his son and daughter were attending university, he enjoyed work more. He was proud of his accomplishments and of the two compounds he helped to discover that were licensed by Virnuc. He was also proud of his long-term friendship with Roman and he spent a good deal of time telling Daniel about their 20-year friendship.

Daniel had done a little reading about Sicily. The island was independent from prehistoric times through the dark ages. It became a part of the Spanish Empire under Aragon in the 1300s. This lasted through the 18th century when it came under the aegis of Piedmont. During that time, the Sicilians defeated the French on several occasions including vanquishing the young Napoleon Bonaparte. Sicily became part of Italy in the late 19th century at the time of the Italian unification. But it maintained itself as a semi-autonomous region of Italy with a different culture according to various authors. Sicily, especially Palermo, was heavily bombed during World War II. Palermo was largely rebuilt with only a few of the older neighborhoods remaining. To this day, Sicilians still maintain a cultural distance from Italy in many ways. The university is one of the oldest in Italy having been in existence since the fifteenth century.

By the time the boat docked it was after 8 pm. The sun was lower but not yet ready to set. They still had an hour drive back to their hotel. But Rico had arranged to stop at a small restaurant on the way back to Palermo where they would have plates of antipasti, the world's best oysters, more wine, and fish in a garlic cream sauce. They sat out on the terrace of the restaurant overlooking the sea and watching the sun set. Daniel was exhausted, but happy.

They were up early and Rico picked them up at 8 am to get them to the University by 8:30. Daniel was given a tour through the laboratories and was introduced to the post-doctoral fellows, students and to the Virnuc employees. In fact, there were maybe thirty young people working there in total of whom ten were Virnuc employees. All but one was a gorgeous, young woman exactly as had been described by Diane back in Cambridge. There was one young man, a Virnuc employee, among all these women. Daniel tried to casually ask Roman and Rico why there were no men in the group. Rico explained that 80-90% of the graduates of the University of Palermo were women and that other than one, he had been unable to hire men. He said that the society on Sicily was very matriarchal. Daniel was prepared to believe Rico's explanations of his all female crew, but he could not shake off feelings that somehow Rico's empire was not all it seemed.

Rico's demeanor had changed overnight. He now reminded Daniel of the senior European professors he had known either through his academic contacts, or more recently, during his sabbatical year in Paris. Rico was King of all he could see. He called the meeting to order. The agenda was to review all of his work, both within his own academic laboratory and within the Virnuc group. In fact, he made no distinction between these groups – all were treated in the same way – with a certain amount of disdain. The one exception was a particularly attractive post-doctoral student, Dorothea, that Rico seemed to highlight with the first scientific presentations of the day. Dorothea was the woman Diane had said was Rico's mistress and student. She presented their ongoing work to identify drugs active against TB, which was the subject of her ongoing work.

From Roman's demeanor and his body language from the very beginning of this daylong meeting, Daniel began to understand that Roman was using Daniel to send a message to Rico. "I've recruited an expert in antibiotics. He will judge your antibiotic projects for me." Daniel started to sweat before the first presentation had been completed.

Rico's problem with screening for anti-bacterial drugs is that he never knew how the drugs might be acting against the bacteria. He just knew that they inhibited their growth. That could either be because they were just general toxins that kill all cells or because they are good antibiotics or anything in between. The problem with this is that even if you find some inhibitor, it is almost always too weak at the beginning to be a real drug. You usually have to modify the molecule chemically to improve its activity and other properties. But if you don't know how the drug is acting and you just work on improving its antibacterial activity, you might also make its human cell killing ability more potent at the same time. It is true that you can watch for this as you improve the molecule, but it is a very risky way forward.

Daniel suggested to Rico that they try and identify the mechanism of action of the drug and then try to optimize based on that specific mechanism. Daniel also pointed out that it was unlikely that a large pharmaceutical company would license an anti-TB drug since the market for such drugs exists in the developing world and not the developed world. It was much more likely to be something that the Global Alliance for TB, a non-profit public-private consortium might license for little or no money. Also, Daniel said, all the public-private consortia prefer drugs that are already in human trials, even if only very early stage trials. So Rico would have to gather enough money from grants to get the drugs into these trials – something that seemed unlikely to Daniel.

Rico was clearly disappointed with Daniel's counsel and continually mumbled that he did not understand why pharmaceutical companies would not or should not be interested in TB drugs. Daniel shrugged his shoulders – it's the way they work and think. But the day was not starting well. Roman was frowning and smirking at the same time.

Rico's team the presented a number of other antibacterial projects where Daniel had some of the same criticisms. This did not improve Rico's mood. Daniel was grateful when the lunch break arrived. Rico left the meeting room "to make some phone calls."

The afternoon was spent reviewing the antiviral projects. Once again, Dorothea was the first presenter. Clearly, Rico was already positioning her to enter into antiviral research after her post-doctoral years. Daniel was wondering when it would be the turn for the Virnuc employees to present their projects. Among Rico's antiviral projects, there were two that attracted the interest of both Roman and Daniel and they both involved inhibitors of HIV, the virus that causes AIDS. They were of novel structure, were reasonably potent and had the promise of being directed at a new binding site on the viral replicase. This enzyme, responsible for viral replication, was the target for most of the anti-HIV drugs already on the market. But a new binding site for a drug on the enzyme might mean that the new drug would be able to kill virus resistant to marketed drugs – the Holy Grail of HIV drug discovery.

One project involved developing one of these drugs as therapy and the other involved using them to prevent HIV transmission to newborns of infected mothers or even to prevent infection of women during intercourse. Roman took over the discussion here and wanted to know the details around the patents that might be protecting these new compounds. The story was complicated and Daniel got a little lost in the details, although Roman seemed to know exactly what was going on. Daniel would learn later that the initial compound was discovered by another Italian professor and that Rico had collaborated with this professor, who was a chemist, to identify more potent compounds. Rico and his collaborator were the patent holders.

The day ended with no presentations by the Virnuc scientists. After the meeting, which had lasted until well after 6 pm, Daniel pulled Lillia and Maria, the lead Virnuc scientist, aside to inquire as to what projects they were working on and why they had not presented anything. They said that their main project was the HIV compounds presented by Dorothea and that the data she presented represented Virnuc data.

"But why didn't you present?"

"Dorothea has not been involved in any of the work, but Rico told us to give her all our data so she could present."

The next day, when Daniel and Roman were on the plane heading to Paris, Daniel asked him to explain the relationship between Rico, Dorothea and Virnuc. Roman just shrugged his shoulders as if to say – hey – its Italy – don't ask. Daniel then broached the subject as to how, practically speaking, Roman expected him to be responsible for a group of Virnuc employees that were working under the de-facto leadership of a consultant and his mistress. He got another shoulder shrug as if to say, that's for you to figure out.

The chemistry group at Paris was a complete relief to Daniel. After Palermo, he expected the worst only to be surprised by how easy it would be to manage Paris. In Paris, Virnuc already had a small business office run by a manager for Europe, Pierre-Phillippe. Pierre was a tall, lanky 45 year old physician who had previously worked for one of the small, French pharmaceutical companies where he was in their project management group for clinical development. At Virnuc, Pierre was responsible for helping to bring in European grant money and for managing budgets, employees, taxes and other aspects unique to the sites in Paris and Palermo. Daniel would have to lean heavily on Pierre in the near future since the budget proposals for the next year were due at the end of September – in just three months.

The head chemist, Frederic Laurent, was a little older than Daniel, of stocky build with glasses and gray, bushy hair. He very much looked like a European professor of chemistry. It became clear to Daniel during a day of meetings that Frederic was one of the shining lights of chemical science in the antiviral world. The contrast between Frederic and Rico could not have been starker. Frederic was brilliant but modest. He was not afraid to state his opinion, but respected the opinions of others including those from his junior faculty and the Virnuc employees working in his lab. Unlike Italy, Frederic had clearly divided the projects in his lab between his own academic projects and those being pursued by the Virnuc chemists under Frederic's direction. The presentations were evenly divided between the two groups with the Virnuc chemists doing their own presentations for the two projects they were pursuing. One was the HIV inhibitor series that Daniel had just heard about in Italy. In Paris, the chemists were showing what they had learned from the previous compounds and how they planned to make new and more active compounds over the next several months. Their second project was to come up with a second generation HCV compound to backup or follow Virnuc's CI-999 that was to go into clinical trials the next year. For both projects, they were able to present detailed and well-justified plans. Daniel's only problem was that they seemed to be working a little bit in the drug discovery dark ages. Daniel was used to using actual structural information using X-rays to define the three dimensional structure of the enzymes and drugs being researched. But Virnuc had no such capability.

Paul Smith, the head of chemistry for Virnuc, met Daniel and Roman in Paris. Paul was tall, balding, and frankly obese, had a full graying beard and was walking with a limp. He had sparkling blue eyes and a not so understated sense of humor. Apparently he had been at his home in England (he commuted home to England from Boston every couple of weeks). He was trying to get his Harley Davidson out of his garage when he ran over his own foot. Daniel took Paul aside and asked him if he would be willing to supervise the Virnuc chemists in Paris using Frederic as a consultant. Paul was very happy to do this since he had been frustrated by his inability to direct the Virnuc chemistry efforts there anyway. Daniel then presented the idea to Roman, and Roman and Daniel together discussed it with Frederic. Frederic was more than happy to change this as he was searching for ways to reduce his administrative responsibilities anyway. So, this was decided quickly and painlessly and would work well throughout Daniel's short stay at

Virnuc. Daniel asked Frederic and Paul to start working with Pierre to put together the budget for the next year. He asked them to include structure work for both the HIV and HCV projects as part of the budget proposal.

In contrast to France, Daniel was at a loss for what to do with Italy. It was clear that managing the Palermo lab from his office in Boston was not going to be easy, and that pulling authority away from Rico was going to be even more difficult. Daniel also worried about the ongoing effect of Rico's affair with his student on the ability of Virnuc to continue pushing their projects forward. The morale among Virnuc's Italian employees was already suffering.

The problem of Palermo would come crashing down on Daniel's head after the budget process. As Daniel was reviewing the budget proposal sent in by Rico, he realized that the cost per employee was much greater, by several times, than that for the employees either in the US or in France. Daniel called Pierre in Paris. Pierre had seen the same thing, but just assumed that this was the way of things in Italy. He never had to worry about it because Roman always approved the budgets from Italy without question anyway. Daniel decided that with the CI-999 IND application due in November, he would save Italy for the next year. Roman approved the budget for Rico with no questions as he had apparently done in previous years. But, when Roman saw the budget for Paris with its component for using structure to drive the chemistry program, he balked. Roman had seen this done in large companies, but he had never really understood the importance of using this approach for antiviral drugs. And he was not about to change his mind. That part of the budget was dead in spite of careful and what Daniel considered to be irrefutable arguments to the contrary. But, Daniel told himself, this is Roman's company, so he said his piece and then dropped the matter. Paul was not quite so forgiving, but like Daniel, he had little choice in the matter.

When January came, Daniel called Pierre.

"I want to audit the Virnuc books in Italy."

"Why?"

"You know why. I need to understand why it is so expensive. They are spending several times more per person in supplies and equipment than any of our other sites. Why? Can you arrange a trip under the guise of a routine science update visit and get Lillia to bring out all the invoices charged to Virnuc for the last year? Also – see if you can get Maria to spend some time with us reviewing the invoices."

"OK. You know that you're playing with fire here. Rico has been Roman's close friend for a long time and he still retains rights to Virnuc's two lead products. And Roman has been approving these budgets no questions asked."

"I know. But if Virnuc ever wants to go public or ever wants to do a deal with a large pharma, everything has to be clean and transparent. I'm no accountant. But if I can see a problem with the Italy budget, so can someone else carefully reviewing Virnuc's books."

Daniel flew over later that month and met Pierre and Rico at the airport in Palermo. They went directly to the lab and were treated to a half-day update on the progress in work on the HIV and HCV projects at Virnuc. Dorothea again led the presentations with little participation from Maria or the other Virnuc scientists. After lunch, Rico excused himself to go to other meetings. Pierre and Daniel said they were going to meet with the Virnuc scientists to discuss personnel evaluations and the budget for the coming year. They adjourned to a small conference room next to Lillia's office where Lillia and Maria had piled all the invoices for the previous year on a table. Maria then went through each invoice and put them into two piles. One was for Virnuc employees' needs and the other were supplies and equipment going directly to Rico's academic lab operation. Several hours later, the pile for Rico was four times the size of that for Virnuc. Rico was clearly inflating his budget request to cover some portion of his academic lab needs every year. Pierre and Daniel looked at each other – now what? Maria and Lillia both shrugged their shoulders. They both knew this was going on and accepted it as part of the deal between Roman and Rico. If this was part of the deal, and Daniel was responsible for the Virnuc operation in Italy, how come he didn't know about it? Pierre knew no more than Daniel and he was responsible for managing the Virnuc accounts in Europe.

Daniel said, "Obviously, I'll have to ask Roman if he knows about this and try and understand what's going on. I've looked at our contract with the University of Palermo and can find nothing about Virnuc being responsible for paying for yearly supplies and equipment for the university. There is just an agreement around lab space and common support like waste disposal, library access and things like that. Virnuc gets that support in trade for a small percentage of revenues from any product coming from the collaboration. There is no mention of anything else." Pierre agreed that that was his understanding of the contract.

Daniel flew back to Boston and alerted Roman that he wanted a meeting to discuss the Virnuc operation in Italy saying that there were aspects that he did not yet understand. Roman was curious, but happy to meet. Daniel called Sally to explain what was going on. He was worried that this would be the proverbial can of worms.

Pierre had prepared a spreadsheet showing the results of their audit of the budget for the Virnuc operation in Italy for the prior year. It was quite clear that 80% of the money was being funneled to the university labs. Daniel explained to Roman the real motivation for his trip to Italy and showed him the result of the audit. He also expressed concern that Dorothea seemed to be the voice for the Virnuc scientists even though she did not work for Virnuc. Rico seemed to have placed her in charge of Virnuc's day-to-day work without consulting anyone – certainly not Daniel. Roman furrowed his brow, looked hard at Daniel, threw the spreadsheet on the conference table and stood up. "This is Italy! What did you expect?"

"If I can find this, so can anyone who looks carefully at our books. If someone from another company wants to look at what's going on at Virnuc Italy – who will present to them – Dorothea – who is not even a Virnuc employee? Rico, our consultant?" The fact that Rico was trying to advance the career of his mistress did not have to be openly discussed.

In fact, Virnuc was in the middle of negotiations to license their key late stage product, teldine, to two different large pharmaceutical companies. Both were known for being straight-laced and highly conservative. Both were looking at all aspects of Virnuc and teldine, including who held the patent rights. At the same time, Roman was still considering an initial public offering to take Virnuc public. He was just waiting for a deal and for the markets to come back up from their low in 2001. Daniel had brought him the kind of mess he did not want to think about or know about. Daniel knew, at that moment, that the person Roman would hold responsible for this mess would be Daniel – not Rico.

"I think we should hire a director to run the site in Italy just like we have Paul Smith running the Virnuc site in France." Daniel said. "Our director would then be responsible for our budget there and for the scientific work of our employees. Rico can go back to being our consultant and Dorothea can go back to being a post-doctoral fellow at the University."

"Rico will raise hell. I'll talk to him. But OK – see if you can find someone."

Even though Roman had agreed to Daniel's plan, he was angry, and not with Rico.

Daniel reviewed the whole situation with Sally that evening. On the one hand, she found Roman to be charming and she liked his family. On the other, through Daniel, she could see the results of his ego. "Can't you just get along and keep your head down like at Penfrel? I like living here and we just got here. But you're going to get us thrown out on our ears!"

"Now that I've opened this Pandora's box, I can't just put the lid back on. What happens happens. I'll do my best, but I'm sure that Rico is not going to react well to a new director taking over his budget. And when Rico explodes, it will be me who gets hit with the shrapnel. I'm really sorry – but its starting to not look so good, and I feel like we just got here."

"That's because we just got here! I love you – try and keep your head down anyway."

Daniel presumed that Roman was willing to agree to this because he was about to close a deal with Gorman Ltd for Virnuc's teldine for hepatitis B. Gorman ended up taking 51% ownership of Virnuc, licensing teldine and acquiring a right of first refusal to all antiviral compounds coming from Virnuc's laboratories. They had to exercise their right after the earliest clinical trials or give up their rights to the compound. With this deal in hand, Roman's next move would be to take Virnuc public. He was anxious to show that he was in control in Italy.

Daniel went about hiring his Director. He put together a job description. Roman approved it with no comment. Daniel asked Pierre to place it in various scientific publications like Science and Nature and to post it on the Virnuc website. A month later, they only had five or six applications from people who were anywhere near qualified. They probably had a hundred or so that were from people finishing their graduate work or post-graduate training and hoping that this could be their first job. Those just went into the delete bin. One of the applicants was Michael Deluca. He was born in Italy where he stayed until the age of three. His parents then moved to Switzerland where Mike grew up and went to school. But when it came to go to university, Mike moved to Montreal, Canada. After studying virology and biochemistry, he went to work at a small pharmaceutical company there – Montreal Pharma. He had been there ten years when he applied for the position at Virnuc. Mike had worked on a number of projects in Hepatitis C and had extensive experience in drug discovery within the pharmaceutical industry. He was ideal. He already had Italian citizenship and spoke fluent Italian, German, French and English. His wife was French Canadian. Mike came and interviewed with the team in Boston including both Paul Smith and Roman. He quickly seduced them both. Paul was delighted when Mike asked why Virnuc was not using structure to advance their programs more quickly. Roman even told Daniel to do everything he could not to lose this guy. As negotiations went on over the next month, Daniel and Roman realized that hiring Mike would require more money than they had initially intended – but Roman was hooked.

Mike was hired. But there was a catch. He would be on a three month probation period depending on how he would get on with Rico in Italy.

Initially, Mike and Rico were the best of buddies. They went sailing together on Rico's boat. They went out to dinner – Mike and his wife and Rico with Dorothea. But Rico was in constant contact with Roman. Daniel never heard from Rico – ever. Mike seemed happy. The Virnuc employees were thrilled. Mike took over their scientific direction and proposed several new approaches to their projects. Even Rico was impressed. But this was to be the calm before the storm. Apparently, Rico thought that Mike would be some sort of figurehead, but that the real director would still be Rico. When Mike took over Dorothea's duties and did not give her any Virnuc work to do, Rico was furious. He called Roman. Roman called Daniel into his office demanding to know what was going on. Roman started on a tirade but Daniel said, "What did you expect? This is what we hired Mike to do. If you think this is bad, wait until Mike submits the budget for next year and wait until you see how this year's budget will be underspent. Rico will go ape-shit."

And he did. Rico essentially closed his laboratory to the Virnuc employees including Mike. They kept what offices they had, but the laboratory was off limits to them. The only way Virnuc projects in Italy could not progress was if Rico's lab took over the work under Dorothea's direction. Virnuc was paying twelve salaries plus benefits for no work. Rico was in breach of contract. And Roman was furious with Daniel for bringing things to this state of affairs. Rico was still speaking to Roman and, in fact, Mike maintained a cordial relationship with Rico. But the Virnuc scientists were not getting back into his labs.

"Our only choice now is to move out of the University. My advice is to get out of Italy altogether and bring the work back here to Boston."

Roman would not hear of the possibility of leaving Italy. First, he warned Daniel that this fiasco would show up in his own performance evaluation. Then he told him to work with Pierre to find a suitable place to build a lab and start getting estimates for cost and time. Daniel assumed again that the pressure of the IPO was forcing Roman to do what was necessary, but never did understand why Virnuc was married to Italy. Maybe Roman was loath to let all those employees go. Daniel even admired him as he reflected that this might be why he was building this new lab in Italy. But several years after Daniel's departure from Virnuc, when the company fell on hard times, Roman shut down the facility in Palermo and all the employees were let go.

Daniel, Pierre and Mike, with help from Maria and Lillia, found a plot of land just outside of town, not far from the university. Foundations for a small building had already been placed, but the project had fallen through. The contractor was looking to replace his lost deal. Roman took over the negotiations and bought the land. But Virnuc had to find a contractor who could build a lab up to Biosafety Level 3 specifications since they would be working with HIV, the virus that causes AIDS. They did, and the contractor promised to have the building up and running within six months. He got it done in a year, but Daniel always thought that the guy lost money on the deal because Roman had forced a heavy penalty clause for late delivery down his throat.

All this happened within one year of Daniel's arrival at Virnuc. Once things started to fall apart in Italy, Daniel was treated to monthly abuse sessions in Roman's office where he would be berated for one thing or another that had gotten Rico upset. Daniel dreaded Rico's regular phone calls to Roman since they always signaled another dressing down. Sally couldn't believe that this was happening. She thought Daniel was a respected scientist in his own right and she firmly believed that Daniel was right in his point of view and that Roman was not. She was furious. She wanted Daniel to resign. Daniel, though, once again, was afraid. He called Bos. "You've got to tell Roman that this cannot continue. If it does – you have to get out of there. I'm really sorry I got you into this. I didn't know that Rico would do anything like this." At one of the verbal abuse sessions, Daniel told Roman that if this continued, his current position at Virnuc would be untenable. He could not continue under these circumstances. Daniel did not think that Roman was even listening.

The deal with Gorman was completed in May, 2003, 11 months after Daniel's arrival. Roman immediately set things in motion for an IPO. He changed the chief financial officer, bringing in someone familiar with the regulatory aspects of going public. The Sarbanes-Oxley Act had passed congress and the new rules for public companies were going to require a large investment for a private company like Virnuc in order to become public. But Virnuc's investors and Roman were determined to go forward. Daniel's own situation did not improve. Roman was spending more and more time away from the office talking to bankers and investors preparing for the IPO. But he still found time to call Daniel on the carpet those times when he was in the office. Daniel decided to try and hold out until after the IPO.

During that year, Virnuc's facility was built and commissioned in Palermo. They finally received all the approvals necessary to open their new BSL-3 lab and start working on HIV and other viruses. Mike Deluca gathered all the employees and invited Rico, Roman and Daniel to a ribbon-cutting ceremony. Maria and Lillia were thrilled to finally be able to get back to work. All the Virnuc scientists were bored with their forced vacations. Rico came with Dorothea and behaved as if everything was normal. He seemed to celebrating the moment with everyone else. To Daniel, it was surreal. At the celebratory dinner that evening, Rico didn't bother to show up.

Daniel was struck by the fact that no one left Virnuc – but he also realized that there could not be so many jobs in Sicily for skilled scientists. In talking to his employees, he realized that no one really wanted to move. The only person who seemed willing to consider working elsewhere was Lillia – but she stuck it out with Virnuc anyway. Even Mike, with all the frustrations of working with Rico and essentially becoming a construction manager for a year, seemed happy to be where he was and anxious to get back to his real job. Daniel knew that given the same situation, Daniel would have been looking to leave as soon as possible.

Daniel and Sally spent many an evening discussing his situation at Virnuc. They both poured over Daniel's contract with the company.

Daniel thought that he could leave Virnuc "with cause" since his job as described in the contract had become untenable and since he felt he had suffered constant abuse at the hands of Roman during the last year. If this were true, Daniel would be entitled to significant financial benefits that he would not get for a simple resignation. Once again, he engaged a lawyer to help at $500 per hour. The lawyer pointed out that, according to the contract, any dispute would have to be settled by arbitration. Daniel would be almost certain to lose under those circumstances. Arbitration almost never favored the aggrieved employee in Lexington, Massachusetts where the arbitration board would be constituted.

A month later, Virnuc became a public company and a week after that, Daniel gave Roman his resignation letter. In it he suggested the possibility that he could resign "with cause." Roman was shocked at first. But then he furrowed his brows as he thought about the statement regarding resignation with cause. Daniel knew that look.

"Haven't you been listening to me for the last year?" He asked. "Haven't you been listening to yourself in your interactions with me over the last year? I did warn you that the situation with Rico was making my job untenable to quote myself. Didn't you hear that?"

"But that situation is resolved now. Rico has essentially stepped aside. We have our own lab and Mike is running it. Why would you leave now? Besides, I need you to keep working on the submission of the dossier for teldine and on the clinical trials for CI-999."

"First of all, I don't see that the situation with Rico will ever be resolved. As long as Rico is there, he will continue to complain to you about Virnuc and you will continue to hammer me about it. Even during the year of lab construction you pounded on me continually knowing that we weren't working with Rico anymore anyway. What has he got on you? This is more than just friendship. What does he have that you need so badly?"

"Rico still holds rights to our patents on both teldine and CI-999. You know that. I have to keep him satisfied that we will hold to our commitments."

"Even when he doesn't have to do that? You could have nullified your deal with him given his breach of contract when he barred Virnuc employees from his lab."

"No – I couldn't. Not with his position on our patents."

"Fine. I feel like Virnuc has made my situation unbearable and my job untenable as I have explained to you in the past. I'm not sure what I will do. I don't want to get into a fight with Virnuc. Is there a way to settle this amicably? Can we come to some sort of arrangement?"

"OK," Roman said holding the letter out for Daniel, "think about a way you could stay to help with our projects without dealing with Rico. Just think about it."

Daniel went home. Sally was waiting for him.

"I don't know what to do. I think he is sincere – but I'm having trouble trusting him."

"After all that abuse, no wonder."

"I was thinking that maybe I could work part time – say 20 hours per week. I would give up my responsibilities for the science, for Boston, Paris and for Palermo. I could keep my responsibilities for manufacturing and for pharmacology and toxicology. I think the piece that Roman is most concerned about is toxicology anyway. There has been some disagreement with Gorman on the interpretation of our toxicology data for teldine and the situation for CI-999 is very complicated. I think that's why he is suddenly playing nice."

"Well, if you do that, don't give up half of your salary and benefits. We would have to sell this house if you did."

Daniel spent the next week talking to Bos and Pierre about his options. He came up with a proposal. He would drop to half time exactly as he and Sally had discussed. But his salary would drop by 25%. Most importantly, he would have the right to sell his stock options over the ensuing year. Since Daniel was an officer of the company, normally, he would not have been allowed to sell his options for at least one year after the IPO. But with his resignation as an officer of Virnuc, he argued that he should be allowed to do so. Roman presented that proposal to

Virnuc's corporate counsel and to the board. They agreed. They offered Daniel a one-year contract under those conditions. Daniel accepted.

Daniel would spend the next year working part time for Virnuc. He was even able to work from home for part of that time. Daniel desperately missed working with the scientists. He continued attending team meetings in his role as head of pharmacology and toxicology but it wasn't the same.

Daniel and Sally had a recurring conversation. Sally would say, "OK. You're going to leave Virnuc. What do you see doing next and where will we be? How about trying to go back to work at a medical school somewhere? You loved working at Cleveland U."

"Well, I've thought about that. I would have to go back as a full professor or perhaps a department chairman. First of all, schools usually reserve those jobs for people who bring a large amount of grant funding with them. I have nothing. I would have to go through a relearning curve in clinical medicine that I'm not looking forward to. And I don't want to get back into the grant rat race again. I've learned so much about drug discovery and drug development over the last ten years; I want to be able to use that knowledge. I want to get back into antibiotic discovery and development somehow."

"I'm sure there are companies who would jump at the chance to hire you."

"I'm not so sure. Between the FDA and the continued abandonment of antibiotics by large pharma, the investment in antibiotics research has all but dried up. The large companies still in the game could get out at any time. I might be taking a job there only to lose it within the year."

"OK. What about small companies? We could stay in the Boston area – there must be lots of small antibiotics companies in the Boston biotech area."

"There are only a few. Most of them have only a single product – usually not such a good one. And their entire life depends on what happens to that one asset. It's risky. I can't think of one offhand where I would like to work."

Daniel and Sally decided to put their house on the market with the idea that he would resign definitively from Virnuc around his third anniversary when his contract would expire. When it came to it, though, it wouldn't happen that way. Daniel and Sally sold their house and put everything in storage. They had rented a furnished apartment in Paris where they would spend the summer while they decided what to do next. Daniel was not ready to retire – he wanted desperately to get back to working on antibiotics where he felt more comfortable and where he felt he had more to offer. But how? He wanted to take a few months to explore his options – away from everything. But when he told Roman his plan, Roman again cajoled him into staying on longer at Virnuc. Daniel agreed on the condition that he could work from Paris. Roman agreed and they parted on good terms. Daniel would stay with Virnuc for another few months and then he would become a consultant for them for another year after that.

Within a few years after Daniel's departure, Roman closed the site in Italy. Virnuc's stock price had sunk to such low levels and the sales of teldine were so meager, that there was not enough revenue for Virnuc to maintain the Italian effort. This was in spite of the fact that Mike and the Italian team had contributed to two new and important projects for Virnuc. They had finally optimized an HIV drug from the project that Dorothea had first presented to Daniel back in 2002. This compound was licensed by Gorman. Mike's team had also identified two backup compounds for CI-999 and both were entering clinical trials.

Paul Smith and Frederic had convinced Roman to allow them to move out of the university in Paris. Frederic wanted to retire from the University and to work for Virnuc on a part time basis. But without Frederic, Virnuc would no longer be able to use the university facilities. This meant that during Daniel's last year or so at Virnuc, there were two building projects occurring simultaneously. The Paris team was cut during Virnuc's hard times, but survived at about a 50% level compared to Virnuc's heyday when Daniel was there. Daniel still runs into them from time to time when he is back in France.

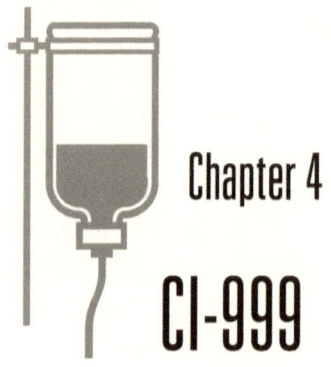

Chapter 4

CI-999

The toxicology reports for CI-999 were going to be late, Daniel's head of toxicology, Jack Harrison, reported. Jack was short, thin and balding and had been Roman's graduate student at Emory University when Roman was a professor there. He had learned toxicology on the fly at Virnuc. Daniel had been impressed with Jack's command of the area especially since it wasn't his initial area of expertise. Jack was just still young and a little inexperienced but Daniel felt he had great potential and was happy with Jack's performance.

Virnuc, since it was a small company (about 50 employees when Daniel arrived), had to contract out most work. The toxicology work was contracted to one of the world's largest contract research organizations specializing in that area. It was also an expensive choice – but Roman believed that top dollar got you top quality. He was right maybe half of the time. In this case, the contractors said they would be unable to get the data through their quality assurance process in time. The entire Investigational New Drug application was waiting for these data and it, like a domino, would also then be delayed. Daniel was having none of it. He called the contractor. Virnuc had been one of their best clients in the recent past. Even though Virnuc was small, it had three large projects all in various phases of testing and all, with one test or another, at this particular contractor. This same contractor had had another recent problem as well. Just before Daniel's arrival, they had run out of drug. Jack had to ship drug on an emergent basis but some animals missed several doses. The entire study had to be revised to account for this slip up. The contractor is supposed to notify the company if they are running out of drug ahead of time so no one misses a dose. Daniel said that he was new at Virnuc, and that if they ever wanted another contract from Virnuc, they would get the data QA'd on time as agreed in their contract. Apparently, between these two problems, they realized that it was not just Virnuc at stake, but their reputation as a reliable contractor. They got the report done in time. The data showed that even the top dose was not toxic to either rats or monkeys. Mary from clinical operations was finally willing to write her section of the IND choosing a starting dose for humans that was 5% of the non-toxic dose seen in the animals. The IND was submitted on time. Daniel got out the champagne that Roman always kept for just such occasions in the lunchroom refrigerator. Within a month, CI-999 was being dosed in human volunteers.

At the same time, Jack had arranged to start chronic toxicology studies for CI-999. It would be dosed over a three-month period to rats and monkeys with the study starting at about the same time as the human volunteer studies. An analysis of data was planned at one month to allow for one-month trials in people. The regulations in the United States allowed one to treat patients for the period of time for which animal toxicity data was available given that the drug was safe in the animals.

Daniel was awakened around 2 am one night with a call from Jack. He was not awake and not prepared for what he heard. "We have monkeys dying."

"What? Did you say monkeys are dying? Are you talking about CI-999? How could that be? The doses in the two-week study showed no effect. These are the same doses and its not even been been two weeks yet, has it? What's going on?"

"I don't know, but half the monkeys in the high dose group are either moribund or dead and the mid-dose group is starting to show signs as well."

"What signs?"

"They are having diarrhea and are off their feed. Sometimes the diarrhea is bloody."

"OK. Stop the high dose group now. Also − give the mid−dose group a rest for a week, then restart at a lower dose."

"Yeah − that's what we had planned. Sorry Daniel − we had no way of knowing this would happen."

"Not your fault. But what the hell is going on with CI-999? Get some of that drug powder back here − lets make sure we're dealing with the same drug. After their recent screw-ups, who knows what is going on with that contractor. I'll call Matt (Virnuc's chief medical officer) and make sure he knows. I'll also try and find out if anything is going on with the human volunteers."

The contractor was dealing with the same drug. The fact that no toxicity was apparent in the earlier two-week study was apparently just chance. If they had used just a slightly higher dose or dosed for another few days, they probably would have seen the same effect. At least that is what Daniel concluded when he saw all the data including the analysis of the drug being administered.

Daniel notified Matt and they met the day after Daniel received the late-night call from Jack. Roman came to the meeting as well. He was worried since Gorman was making it clear in their negotiations that they valued CI-999 − not Virnuc's more advanced teldine. The HCV market was much more attractive than the hepatitis B market. Matt said he would put a so-called stopping rule in the protocol they were using for the volunteers that would stop dosing if someone started to have abdominal pain or loose stools. Roman said he would deal with Gorman. Daniel asked Matt for a monthly update on the clinical trials in volunteers. Matt agreed − but these updates never appeared. The only way Daniel could get data on the trials was to attend the monthly clinical team meetings where the ongoing studies were discussed.

In fact, the clinical trial in volunteers went very well. During a two-week dosing period, even at the highest dose used, there were no side effects beyond those seen in patients receiving sugar pills. Of course, the trial was small with each dose group having six people getting drug and two getting sugar pills. If there was a small effect of the drug, you might not expect to see it in such a small trial. Or the opposite could be true. A small effect might be magnified just by chance when using such small numbers. But that was the risk you accepted.

In the meantime, the chronic toxicology study had progressed for over a month. Daniel and Jack had planned an analysis at that time. When the pathologists examined the intestines of the monkeys who had died earlier, those who were sick and even some of those who displayed no apparent symptoms, they found lesions that looked like the ones you would see after dosing toxic cancer drugs. The cancer drugs prevent cells from dividing, so rapidly dividing normal cells like those in the bone marrow and intestinal tract are the one most affected. It looked like CI-999 was toxic to intestinal cells just like some cancer drugs, but it did not seem to affect other cells like those in the bone marrow. Jack and Daniel had definitely identified the end-organ toxicity associated with CI-999, at least in monkeys. Rats continued to tolerate CI-999 very well.

Since the regulations allowed for equal time between animal and human dosing, after one month of dosing in the animals, human trials could continue through one month of dosing. The clinical plan was to carry out a small trial in HCV-infected patients, using three different doses of CI-999. The low dose would be just below that calculated to have a therapeutic effect. The high dose would be three times the dose needed for a beneficial effect and the mid-dose would be two times that dose. The volunteers, in this case, would be chronically infected with HCV, but would consist of those with no liver damage and therefore those who did not really need effective therapy at this point. One of the concerns was that if people were treated with CI-999 alone, the mutation rate of the virus would lead to resistance to CI-999. But since there was an effective, if toxic, therapy already available, and if the patients eventually needed therapy, there would be no problem going ahead even if they had received a month of CI-999 at some point. The problem was that the one-month tox data from the monkeys would only show that the mid-dose in humans could be justified since the high dose was similar to that where the monkeys had symptoms. Virnuc argued with the FDA that since they now knew the target organ, the intestine, for CI-999 toxicity, they could monitor the patients and that people were not necessarily monkeys. They could be more like rats that tolerated CI-999 very well. The FDA agreed and allowed the trial to proceed.

In this trial, CI-999 showed a clear dose effect on knocking down hepatitis C virus in the blood of infected patients. The low dose showed a minimal effect. The mid dose had a greater effect and the high dose knocked the virus down by more than ten-fold within one month. One problem was that there were a couple of episodes of abdominal cramping in the high dose group – but the patients were able to take the drug for the entire month and never were severely ill enough to reach the stopping rule that Matt had put in place. Everyone was thrilled with these data – except Daniel. Matt and Roman brought out the champagne. Gorman purchased a majority interest in Virnuc. Virnuc could launch its IPO. But Daniel warned that they would probably have to use the high dose in larger trials. These trials would involve patients who were ill with their HCV infection and they would have to take drug for anywhere from six months to two years. Daniel felt that the abdominal pain seen during the one-month treatment of these infected but healthy patients could well be a harbinger of things to come.

Roman and Matt were undeterred. Apparently Gorman agreed with their enthusiasm. Virnuc submitted a protocol to the FDA where they planned to treat patients with standard of care therapy alone, in this case interferon plus ribavirin, or that therapy plus CI-999. CI-999 would be used in a low dose and a high dose as in the earlier one-month therapy. They would follow the amount of virus in the bloodstream of patients during a six-month treatment period. At that point, those patients who had undetectable virus in their blood would discontinue therapy. Six months after that, if they still had no detectable virus, they would be declared to be cured. Stopping rules for abdominal pain or diarrhea were also included in the protocol.

Daniel once again asked to be kept informed of the safety data from the trial on a monthly basis. All this occurred when he became a part-time employee. Not only did he not get reports from the trial, but he was also disinvited from attending the clinical team meetings. Daniel objected, but neither Roman nor Matt was listening. The three-month tox data, in the meantime, came in. It showed that for rats, there was a three-fold safety window for the high dose in humans. That is, rats showed no toxic effect of CI-999 at three times the highest dose used in humans. But monkeys were much more susceptible to the toxic effect of CI-999 and they had toxic effects at one-half the high dose being used in the human trials. When Daniel finally resigned from Virnuc, Roman asked him to stay on as a consultant to help with the submission of the teldine dossier for market approval and to help get CI-999 into later stage trials.

One month later, Daniel was in Washington, DC. Virnuc and Gorman were meeting to rehearse their presentation to the FDA. This was supposed to be the meeting with the FDA that would allow Virnuc to perform the final trials required to get approval to market CI-999. All the data from the phase 2 trials would be reviewed with the FDA and Virnuc's proposal for phase 3 trials would be presented. Daniel was asked to attend the meeting in case the FDA had questions on the animal toxicology studies that had been performed. At the rehearsal, Matt presented the data from the early trials in healthy volunteers, from those in healthy but HCV-infected volunteers, and in HCV patients where CI-999 was added to standard of care therapy for patients suffering from HCV infection. Daniel had never seen any of the latter data and he was looking for the data he knew Matt would show – the safety data from the trial. The standard of care, interferon plus ribavirin, was toxic itself. It caused fever, nausea, loss of appetite and anemia – but with prolonged administration it cured almost half of HCV patients. When the single slide with one table on safety finally appeared, Daniel jumped out of his seat. The data clearly showed an increase in severe abdominal pain and diarrhea and much more discontinuation of therapy due to these side effects in the group receiving the high dose of CI-999 compared to standard of care alone. The trial data also showed that the low dose did not clear virus infection any faster than standard of care alone, whereas the high dose showed a major benefit with 80% of patients reaching undetectable virus levels at six months. So, the high dose was toxic but effective. The low dose was not toxic but was ineffective. Daniel said, "Matt, don't you think that the FDA is going to want to discuss the safety data in more detail than just this one slide?"

"No. This discussion will center on whether CI-999 works against the virus or not and at what dose. We're proposing going forward with the high dose for our final trials and for our marketed dose."

Just then, the Virnuc statistician, Maggie Lin, got up and said, "But Matt, you know that for the last two weeks the FDA has been requesting a variety of safety tables from our study. Maybe Daniel is right."

Matt was adamant saying that CI-999 provided a substantial benefit to patients with negligible safety risk and that the single slide would suffice for the FDA. The Gorman representatives said nothing.

At 8 am the next morning, Daniel found himself in a conference room at FDA's new facility in White Oak, Maryland. The head of the FDA office of antiviral drugs, Gloria Morrison, was seated at the head of a long, oval table. She was joined by FDA staffers – all on the left side of the table. The right side of the table was obviously for the Virnuc and Gorman team. Matt sat next to Gloria. Daniel sat at the far end of the table next to Maggie. Gloria welcomed the Virnuc/ Gorman team to the FDA. She started the meeting by having people around the table introduce themselves. Then she said, "We do not consider this an end-of-phase-2-meeting, but rather a clinical update on Virnuc's CI-999. We would like to start the meeting by showing you our own analysis of the Virnuc trial data that you have submitted to us." Beyond the introductions, the Virnuc/Gorman team said basically nothing for the rest of the meeting. Gloria and her team went into great detail comparing the toxicity of CI-999 with its efficacy during the six months. They were not satisfied that, given the number of patients who had to discontinue CI-999 therapy, the benefit was worth the risk. She made reference to the monkey data and concluded that people were more like monkeys than rats and that with more prolonged therapy, which would probably be required to achieve a higher cure rate than interferon plus ribavirin, the risk to patients might be substantial. She ended the meeting by asking Virnuc for more data on more prolonged therapy or to explore other doses in between the two doses used in an effort to identify a safe but effective dose. Daniel left quietly saying nothing. But he was angry. He was angry with Matt for being so unrealistic and with Roman for letting things come to this point.

Virnuc was already a publicly held company. With the announcement of the results of this meeting with the FDA, Virnuc's stock plunged more than 50%. Virnuc struggled to find an effective but safe dose for CI-999, but when they applied to the FDA to go forward with the dose they finally chose, the FDA refused saying it was unlikely to be beneficial even if it would be safe at that dose. And that was that for CI-999.

Part Three

The Next Thing

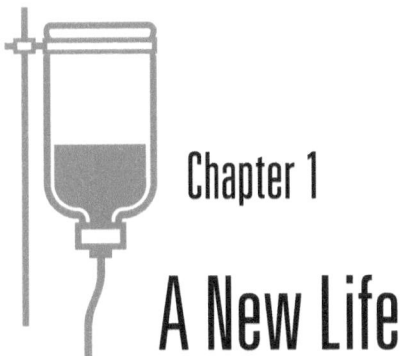

Chapter 1

A New Life

Daniel and Sally found an apartment in the 11th arrondisement of Paris that rented for about 2500 Euro or $3250 per month. They moved in towards the end of June 2005. The apartment was on the 6th and last floor (7th floor for Americans). It had been three maid's rooms (*chambres de bonne*) in the previous century. Someone had knocked down all the walls and joined the three tiny one-room apartments together. It had a large living-room-kitchen area (*cuisine Américaine* according to the French) with floor to ceiling windows looking over the roofs of Paris to the Eiffel Tower. There were two bedrooms with one just off the living room at the front and the other near the toilet at the back of the apartment. The rear of the apartment faced east and got morning sun while the front saw a glorious sunset almost every day. The one problem was that the apartment was warm with all the sun filtering in and without any air conditioning. Daniel and Sally had to close the curtains in the front room during the hottest part of the afternoon every day. Otherwise, the apartment became intolerable even with the powerful fan they had purchased.

The floors were all wood and Daniel and Sally removed their shoes so as not to disturb their neighbors below. They also bought a few cheap throw rugs to help lessen the noise. The rear of the apartment faced a small courtyard, and the odors from the kitchens of the apartments below would waft up through the open windows every evening. By August, Daniel and Sally were among the few tenants staying in Paris. The city was emptying for the annual vacation. But it was not like it had been when they lived in Paris ten years previously. Since that time, France had limited the workweek to 35 hours. This essentially gave French workers an extra two weeks of vacation per year. Many would spread this time out throughout the year so that there were more people and more traffic in August of 2005 compared to 1995.

While he was still at Virnuc, Daniel was placed on a task force for the Infectious Diseases Society of America. Daniel had been active in the society almost since starting his job at the VA hospital in Cleveland 25 years previously. He had served on a number of committees and knew the officers of the society very well. The task force was focused on improving the pipeline of antibiotics in development since the FDA was approving precious few new antibiotics and there were very few in development heading for approval. The physician members and officers of the society saw that, as resistance rates among bacterial pathogens were rising, they might face a day when for some patients, they no longer had effective antibiotics. In fact, many society members were already seeing at least a few such patients every year. The task force was to develop a white paper on this topic providing suggestions of ways to open the pipeline for new antibiotics. The task force had its first meeting in the spring of 2003. Daniel stood strongly for reforming the FDA. He believed that the FDA's insistence on increasing stringency and therefore the size of clinical trials for antibiotics had sent a chill through the pharmaceutical industry. By 2003, many large pharmaceutical companies like Lilly, Roche, Bristol-Myers Squibb, Wyeth, Penfrel and others had openly abandoned the antibiotic research area and fired the scientists involved in that work. When combined with the constant mergers in the industry, there were very few large pharmaceutical companies left doing antibiotic research. The industry thought that the FDA was hostile to antibiotic development, that their return on investment would be marginal, especially given the increasing costs of trials being forced on them by the FDA, and many were just plain discouraged by the scientific opportunity since they had failed to find new antibiotics in spite of years of searching. Daniel thought that the first step to bringing industry back to the area would require FDA reform.

The other stance that Daniel took forcefully was to provide a monetary incentive to pharmaceutical companies that would be large enough and attractive enough that they would be assured that they would get a return on their investment. Daniel thought that a wild card patent exclusivity would be sure to do the trick. The wild card patent exclusivity would allow a company, for example Pfizer, if they successfully developed a new antibiotic active against infections cause by highly resistant bacteria, to gain an extra 6-24 months of exclusivity on a product from their portfolio of their choice. In the case of Pfizer, they might have chosen Lipitor, their $15 billion dollar-a-year drug to reduce cholesterol and prevent heart disease. There is no way an antibiotic would ever be able to achieve sales like that. And – the revenue from such a choice would fund antibiotic research and development at the company for years to come while providing a comfortable and guaranteed return on their investment in antibiotics.

While the idea of reforming the FDA was something the entire task force could get behind, Daniel ran into stiff opposition from some members on the wild card patent idea. The generics industry would fight this tooth and nail, they said. The person most opposed to this was a staff member at the Society who was responsible for the Society's lobbying efforts in congress. He made some initial inquiries on the hill and found a lot of closed doors. He said that the proposal would be dead on arrival in congress. Not only would congressmen and senators have to deal with the wrath of the generics industry, the pharmaceutical companies were so unpopular that no one wanted to be associated with giving them money that would come directly from consumers pocketbooks in the form of delayed generic drugs. They accused Daniel of being a little like Don Quixote. But Daniel Quixote would not give up on the idea. He knew it would work and it was hard for him to see anything else at the time that would. All the other financial incentives being discussed would never attract large pharma back into antibiotics. These other incentives included things like extended patent exclusivity for the new antibiotic. But that would never work because when the bean counters at companies discounted for inflation, those last years where the additional exclusivity might come into play counted for nothing.

Another incentive discussed was tax credits. This might work for small companies, but would never be enough to attract the large companies back to antibiotics in Daniel's opinion. Daniel's key suggestions were finally included in the white paper that was published a year after the task force's first meeting. FDA reform would only occur eight years later and the wild card was, as the Society had predicted, DOA.

Many in academia and in the pharmaceutical industry admired Daniel's work on the white paper. His successful struggle with the FDA while he was at Penfrel also gained him a certain reputation. He was invited to give lectures on these topics in the US and around the world. Roman was happy to have Daniel do this as it reflected well on Virnuc and he still had not given up on the idea of developing a new antibiotic.

While in Paris, Daniel kept up small amounts of work for Virnuc. He was helping finalize the toxicology reports for both teldine and CI-999. But this did not consume much of his time and he was free to relax and think about the future.

Sometime in July, Daniel received a call from one of Virnuc's board members, Deborah Cranston-Smith. Deborah was a partner in a venture capital firm, Mitsubishi Ventures, working out of Mitsubishi's offices in London. They were an early investor in Virnuc and she remained on the board after Virnuc became a public company. They had met several times, but they did not know each other well at all.

"Hi Daniel. How are you doing? Are you enjoying Paris? We have a place in the South of France and we'll be heading there in another month or so. We should get together."

"Hi Deborah. We're good and we like Paris. We needed this." Daniel didn't think that Deborah would call just to invite him down to her place on the Cote d'Azur near Cannes – and he wasn't sure he wanted to go anyway.

"I'm calling because we are considering an investment in an antibiotics company. This company would be a spin-out from a large pharma. Would you be willing to help us analyze the opportunity and, if it looks attractive, help us structure the spin-out?"

Daniel needed no convincing. He agreed to take a trip on the Eurostar to meet with Deborah's team in London to hear more about the proposed spin off. But before he could leave on that trip, he got another call. This time it was from Roman.

"We're thinking of buying a small antibiotics company here in Boston. When you come back from Paris, would you be able to help us evaluate it? They have a product that is about to start clinical trials and I think you will have special expertise in this particular antibiotic." He wouldn't say more. Daniel reminded him that he would be resigning from Virnuc in September, but said he would be happy to help Virnuc as a consultant after that.

Within a month, he received yet another call. This time it was from Gorman. They wanted Daniel to be part of their antibiotics scientific advisory board. They proposed one or two meetings per year with the option to call him for advice at other times as well. Daniel pointed out that he would still be at Virnuc until September, but after his departure, he would be happy to work with them.

Daniel's new career seemed to be arriving without him doing anything. But there were many potential problems with the idea of becoming a consultant. Would there be enough work to keep him busy? Would it be challenging? Could he actually get into the details of the science and the clinical development strategy he liked so much by just being a consultant? Would he ever be able to "own" a project the way he did at Penfrel? Would he be able to step back and keep his mouth shut when needed?

Daniel was a fighter. When he felt strongly about something – he went to the mat. His struggle at Penfrel with the FDA was emblematic of that aspect of his personality. But he knew that would not necessarily work for a consultant. He would have to make his case, maybe more than once and maybe forcefully – but he would always have to know when to step back. It would never be his project and never be his company. At the same time though, would he be able to take some ownership of the product or the project at hand? Could he take some credit for getting molecules along the path to market as a consultant?

He called Bos to discuss the whole idea and to air all of these questions. Bos had been consulting for a long time – not as his primary business – but he certainly would know the answers to these queries. Not really. Bos himself was not one to step back and that had limited his consulting activities to a certain extent. Also, since consulting was not Bos' primary job, he never could get into any of the projects in enough detail to really take credit for anything.

Daniel knew another physician who had left industry to become a consultant – Henry Maloon. Henry had worked at two large pharmaceutical companies and had been involved in the design and execution of several late stage clinical trials that led to successful applications to market two antibiotics. He then became a consultant and was doing very well. Henry was busy, maybe too busy, and was clearly enjoying his life. Henry had taken "leave" from his consulting business on at least two occasions to join companies as their chief medical officer to lead the clinical development of their products. Of course at that point he definitely owned those projects. Henry provided a good deal of sage advice for Daniel. But in a number of areas, he would remain silent. How do you decide about fees to charge? Should it be hourly? Should he accept stock options from small companies in lieu of money? Henry thought it was inappropriate to discuss any of that. "You understand, don't you?" he asked. Not really, Daniel thought to himself.

Daniel didn't know the answer to any of these questions. But he had dealt with quite a few consultants during his years at Penfrel and then later at Virnuc. In thinking about becoming a consultant, he decided that he would never accept stock options in lieu of an hourly rate, but would accept both under the right circumstances. He set an initial hourly rate based on what he had been paying consultants in the past. He provided for volume discounts – with a cheaper rate for an entire day of consulting. He also decided to charge for travel time since it looked like some of his clients were going to be in Europe and that was going to mean significant time on planes.

Another very attractive feature of being a consultant was that he and Sally could live wherever they wanted. Daniel argued for Northern California – like the Sonoma valley. He had become quite a wine freak during his time in France. He and Sally had even taken a wine tasting course at a neighborhood wine store during one of their sabbatical years in Paris. But Sally voted for New England. Both of their daughters and two grandchildren were in the Boston area. "Why would we move out to California? Its far from Boston and farther from Europe." So – New England it was to be.

"Is this consulting gig going to work for us? Financially, I mean." Sally asked.

"I think that we have enough put aside to retire if we really have no choice. Of course, we'll either live more modestly or live well but not very long. At least consulting should give us a little pocket money for a few years."

At the end of September, Daniel and Sally returned to Boston to stay with their youngest daughter while they found a place to live. Daniel finished up with Virnuc and signed a consulting contract with them. His first task would be the end-of –phase-2 meetings on CI-999. His next one for them would be the antibiotic company in Boston.

Sally took on the task of scouting for places to live. They didn't want to sink too much money into a house since they didn't know exactly how this consulting gig would work out financially and they were still young for retirement. Sally looked farther and farther away from Boston. Once they got into western Rhode Island and Eastern Connecticut, prices came down to manageable levels. Sally settled on a few towns in Southeastern Connecticut and made several trips there while Daniel started working from their daughter's apartment. Sally narrowed the choices to about five possibilities. Daniel went with Sally one week to check out the area and the houses Sally had prioritized. On looking at the houses online, Daniel focused on one in particular and said "This is it," just based on the photos and the description. That was the house they bought. They moved in early November of 2005.

Their move was exactly what they had envisioned. They saw daughters and grandchildren regularly. Sally had her garden. They were living in the country and just a mile from Long Island Sound. They heard foghorns and could see the ocean from their terrace (in the winter when there were no leaves on the trees). Sometimes they could see all the way out to Fisher's Island. Boston was 90 minutes away by car and New York was more like 3 hours away.

Daniel was able to get back to two activities he had missed sorely. When he was in high school and in college, he was an avid guitar player. He played classical and quite a bit of flamenco. In those days, he would entertain at cafés and bars and was even featured once on a public television concert. He had dropped the guitar while in medical school. But when the moved to Connecticut, he picked up his old guitar again and started relearning. He found a classical guitarist in his town and started taking lessons again. His repertoire expanded quickly and he played as much as he could in and around his travel schedule which was becoming hectic.

He also went back to fishing. Daniel used to fish with his grandfather in Florida. During their younger years, when they could afford nothing else, Daniel and Sally would take the girls on camping vacations where, for Daniel, fishing was a major activity. The move to Connecticut allowed Daniel to get back to fishing on the ocean and on the nearby streams for trout. Life was good – mostly.

Daniel once again hired lawyers. This time, he needed help setting up a business. He had a business plan. It was pretty simple – it was to be a single employee sole proprietor company named Antibiotics LLC. He got a corporate credit card and a bank account for his business. He was almost ready to go.

Daniel also had devised an important policy – a conflict of interest statement. This policy stated that Daniel would only consider that there would be a conflict of interest between clients if both the chemical structure of the compounds involved were similar and the bacterial spectrum – that is the kinds bacteria killed by the compounds –was similar. This was important because many antibiotics targeted the same sorts of bacteria. If each one of those put Daniel in conflict – he would only ever have a single client. On the other hand, chemical structures were similar to each other much less commonly although it did occur. Over Daniel's subsequent eight years of consulting, he only had to invoke his conflict of interest policy one time, although several clients asked him to provide the policy before they would engage him.

By the time his consulting business was formally established, Daniel already had a number of clients. Based on the volume of calls he was getting, he thought that number would increase quickly. His clients ranged from investors seeking advice on various investments with newly formed companies working on antibiotics, to small companies trying to advance their products from the lab into clinical development to tasks from Virnuc. Daniel thought about setting up a website or doing other marketing, but he was so busy from the outset, that he dropped that idea very quickly. By January of 2006, he already had ten regular clients plus occasional investor evaluations to do and was traveling regularly. At this beginning of his consulting career, he had no large pharmaceutical companies as clients. Daniel was chagrined that this was true but he was consoled by reminding himself that there were so few still involved in antibiotics research and that they all thought they knew everything anyway.

Chapter 2

Daniel and the FDA

During his years at Virnuc, Daniel kept up with the antibiotic area. He continued to participate in the antibiotics working group of PhRMA, the pharmaceutical companies' trade association. Through them, he learned that the FDA had returned to insisting on more stringent trial designs for antibiotics after teracil. Once he left Virnuc, he was no longer able to participate in the working group, but he kept abreast of developments through his contacts both at FDA and through his clients.

In 2006, antibiotic development unraveled at the FDA and then for the industry. Daniel, at the beginning of his consulting business, was worried that his clients would go out of business. If that happened, Daniel's business was doomed from the outset. The FDA could freeze investor confidence and could force more companies to abandon antibiotic R&D. In point of fact, Daniel's clients stayed their course, but for many years further investment in antibiotic R&D was frozen and several more companies like Pfizer and Johnson & Johnson stopped their antibiotics research and development and fired most of the employees they had working in the area.

It all started with an article that appeared in the Annals of Internal Medicine in January of 2006. The article described three cases of severe liver toxicity that seemed to be caused by an antibiotic that the FDA had approved two years previously, telithromycin or Ketek. Ketek was developed by Sanofi-Aventis. This was an important antibiotic at the time because the key bacterial pathogen causing respiratory infections like pneumonia, sinusitis and bronchitis was becoming highly resistant to drugs like azithromycin (the Z-pack) and erythromycin. In fact, 35% of such strains were resistant in the US alone. Ketek was related to these drugs, but remained active against the resistant strains. For patients allergic to penicillin or for children who could not be treated with the other key class of antibiotics, the quinolones, Ketek could be an important alternative therapy.

It completed its clinical trials in 2001 and was submitted to both Europe and the FDA. Europe approved Ketek for use in bronchitis, sinusitis and pneumonia. The FDA balked. They were worried about possible arrhythmias of the heart based on findings in animals. They asked Sanofi-Aventis to carry out a large study in humans, either volunteers or patients, to establish the potential of Ketek to cause cardiac problems. Sanofi-Aventis did just that. They carried out a study of 24,000 subjects in a single year. This would be the largest antibiotic trial in world history and was carried out in record time with a record number of clinical centers participating.

But the study was too ambitious for the company. The physician leading the study at the center that enrolled the highest number of patients was convicted of fraud for inventing patients and data that was then submitted by Sanofi-Aventis to the FDA. Another investigator was arrested for drug use and threatening the arresting officers with a pistol hidden in his underwear. The FDA found the study so badly flawed that they refused to use the data. But by 2004, Ketek had been sold in Europe for several years. The FDA examined its safety history in thousands of patients treated throughout Europe and based on the lack of any significant findings, approved the drug. In retrospect, this was extremely controversial since those safety data are just based on voluntary submissions by physicians or patients. These data are not controlled in any way and therefore remain somewhat suspect.

Then there was the article in January 2006. Shortly after the appearance of this article, the FDA convened an advisory committee to review

Ketek once again. This would be the third time for Ketek. Prior to this advisory committee meeting, several disgruntled FDA staffers who had worked on the original Ketek submission leaked emails demonstrating their opposition to approving Ketek on the basis of post-market safety data from Europe. These emails were published in the New England Journal of Medicine and the New York Times, Washington Post and other papers. They found their way to the desks of key congressmen.

But at the advisory committee meeting, based on the published article plus a review of all available safety data on Ketek both in the US, in Europe and from key databases from the Kaiser Foundation and others, there seemed no doubt that Ketek could, rarely, cause severe liver toxicity. But it appeared to occur during 1 in 100,00 to 1 in 200,000 courses of therapy. When the FDA looked at their own safety data on other antibiotics, Ketek was really no different than several others including a number of generic drugs. It appeared to be safer than Tylenol.

In spite of FDA's own data, though, the FDA was determined to kill Ketek. Senator Grassley and Congressmen Markey had both threatened congressional investigations of the approval process for Ketek. The FDA staffers spent months of days and nights and weekends responding to inquiries from these congressional offices. To kill the drug, the FDA relied on several previous advisory committee meetings where they had been exploring the benefits of antibiotics to treat ear infections, sinusitis and bronchitis. The committee and the FDA believed that these were "mild" infections and that the data supporting the idea that antibiotics actually work to provide a clinical benefit in these infections were at best controversial. So, they reasoned, if there is no clear benefit, then the benefit risk ratio is by definition zero. The committee agreed. Approval for use of Ketek in sinusitis and bronchitis was withdrawn, but its use for pneumonia was allowed to continue.

Daniel attended the advisory committee meeting and was appalled at the way Ketek was indiscriminately ground to dust. He stood up during the public commentary section of the meeting and asked the FDA,

"What are you going to do about all the generic antibiotics where your own data indicates that they are as toxic as Ketek, but which remain approved for sinusitis, bronchitis and ear infections? Shouldn't they, too, have their approvals for these indications withdrawn?" The FDA responded by saying that they would look at the generics but that the task would be "like climbing Mt. Everest." Daniel never understood that. But to this day, the FDA's Everest remains unconquered. Daniel and others wrote editorials decrying the FDA's decision as not having been based on their own data.

Shortly after the Ketek advisory committee meeting and the subsequent withdrawal of most of Ketek's marketing approvals in the US, the FDA made a number of startling announcements. First they said that all previous guidance on how to study antibiotics in clinical trials was hereby rescinded. Next they said that any new antibiotics to be studied for market approval would have to justify the non-inferiority margins (the statistical difference between the new drug and the comparator drug) proposed for their trials. But no one outside the FDA understood exactly what was meant by "justify." Finally, they said that new antibiotics for treatment of ear infections, sinusitis and bronchitis would only be approved after having shown that they were both effective and safe in so-called placebo-controlled trials.

Finally, the FDA reorganized its entire antibiotics section. The head of the section that had approved Ketek using the unprecedented approach of using post-market safety data from Europe to do so was reassigned to vaccines and ultimately was sent overseas. Those who leaked emails found jobs outside the FDA where they continued to have influence in congress in criticizing the industry approach to antibiotic development.

Daniel, based on his attendance at several of the FDA's advisory committee meetings, had anticipated much of what FDA had now formalized into regulation. But he was devastated. He knew that the lack of clear guidance on how antibiotics could be approved in the US would only lead more companies to abandon the area. If you don't even know how to proceed, how could you know how much you had to invest and whether you would ever get any return? Uncertainty is the enemy of progress in the pharmaceutical industry.

He also thought that the FDA's new requirement for placebo-controlled trials in the treatment of bronchitis, sinusitis and ear infections would mean the end of any new antibiotics designed for those infections. Daniel tried to imagine the situation for his daughters. He remembered waking up with them at night when they would have fever and a severe earache. This would lead either to the emergency room, to Daniel writing his own prescription for antibiotics for them, or to a quick visit to their pediatrician or some combination of all of those. He could not imagine a setting where the emergency room doctor would say – "Hey – I have a new clinical trial I would like you to enroll your daughter in. In this trial, 50% of the patients would get an antibiotic that we think will work and the other 50% would get a sugar pill. How about it?"

Actually, two such trials carefully defining children with true infections of the middle ear were conducted in later years. They took many years to enroll just 100-200 patients. The studies showed definitively that antibiotics worked in these infections and that the children who got the sugar pills were more likely to end up with serious complications like meningitis and infection of the bone around the ear canal and perforated eardrums. The FDA would finally give up this ghost after the data from those two trials were published. But in 2006, the FDA was adamant.

Daniel knew that the FDA's policy would also, with a single stroke of the pen, wipe out one of the most lucrative markets for antibiotics off the face of the earth. It was the double whammy - Uncertainty for development, and a guaranteed loss of markets.

Daniel confronted the FDA at every opportunity. He spoke publicly at advisory committee meetings. He wrote a book and started a blog to discuss the challenges of discovering and developing new antibiotics active against resistant bacteria. He tried to cajole the FDA into being more open about the fact that even if the role of antibiotics in so-called mild infections was controversial, that fact meant that there were more points of view than just the one that the FDA espoused. In fact, the FDA admitted that the available data showed that antibiotics prevented death (as in death!) from the most severe cases of bronchitis. They were just not so sure about its role in more mild cases of bronchitis. For sinusitis – foggedaboutid!

Daniel was comforted by the fact that the FDA seemed to still allow trials to go forward in serious infections in hospitalized patients as opposed to "mild" infections in non-hospitalized folks. But that would not last long. Daniel was attending a Gordon Research Conference meeting in 2007 – not long after the Ketek advisory committee meeting and the release of the new guidance on mild infections. There was a session towards the end of the meeting where there were presentations from industry, from the infectious diseases society and from the FDA. The director of the antibiotics group at the FDA was their speaker. He announced, in a quite matter of fact manner, that the FDA no longer knew how to define the non-inferiority margin for trials in skin infections. This would mean that trials where new antibiotics were being studied for the treatment of these infections in seriously ill hospitalized patients were suddenly thrown into doubt. Would the FDA, who had agreed that the trials were adequately designed a year earlier, now renege on their word?

The conference turned into pandemonium. Each such trial costs a company in the realm of $30 million. You normally had to do two such trials to get a drug registered for a total cost of $60-70 million. Some of these trials were well along their way with most of the money already spent. Daniel had a client in just this position at the time. Someone yelled, "What about ongoing trials? Surely you will stick to your given word." The Director yelled back, "We reserve the right to adjust our requirements to the changing science of clinical development."

Daniel was sick. In his mind, there was no changing science of clinical development, just a changing attitude of an FDA gone mad with fear of congressional investigations and loss of jobs. Daniel's client, with

Daniel's help, desperately worked to try and justify their trial design according to what the FDA wanted. But since no one, including the FDA, knew what the FDA wanted, his client finally had to stop their trials. They could no longer be sure that the FDA would accept any resulting data as meeting their requirements for market approval. Several other companies were caught in the same position. At least one went belly up because of the loss of investment in a trial started prior to 2007 and their inability to obtain additional funding for other trials given the sudden lack of enthusiasm for antibiotics as an investment opportunity.

Daniel's meetings with investors reflected the new reality. He was suddenly faced with individual investors who were still interested in antibiotics, but who were unable to convince their management to invest no matter how potentially attractive the opportunity might have been. For investors, antibiotics joined diabetes and weight-loss as areas where new investment would be *verboten*. They simply felt that the regulatory uncertainty created by the FDA was too big of a speed bump to overcome in the near term.

Within a year, the FDA and the Foundation for the National Institutes of Health under the leadership of Henry Maloon had found a way to justify a non-inferiority margin that would allow trials of skin infections to once again go forward. Daniel was not directly involved in this effort and could only watch from the sidelines. The scientists on the foundation and the FDA identified two clinical trials of sulfonamide antibiotics used for a very specific kind of skin infection that were carried out back in 1937. The antibiotic was compared to UV light therapy. The FDA accepted UV light as a placebo for their purposes. Streptococci most commonly caused this particular type of infection while today's infections are usually caused by more difficult staphylococci. The methods back in 1937 were nothing like the methods we use today to design and conduct trials. But it was all the FDA could find and all they would discuss. The 1937 trials measured not cure of infection, but whether the lesion of the skin stopped progressing or not within the first few days of treatment. In these trials, sulfonamide antibiotics stopped lesion progression in 98-99% of patients compared to about 75%

for UV light within the first 48 hours. This demonstrated a treatment effect of 23-25%. So, the FDA decided that all trials of antibiotics for skin infection would have to show a halt to lesion progression within the first 48 hours.

When Daniel heard this news, he was both overjoyed and angry at the same time. He was overjoyed because at least companies would be able to carry out the trials now required by the FDA. He was angry because no one, as in NO ONE, cared about halting the spread of the skin lesion as an actual endpoint. The only endpoint patients and physicians, to say nothing of payers like insurance companies, cared about was whether or not the patient is cured. Of course, physicians examine their patients every day to determine if the skin lesion in progressing or not. But to make that alone the endpoint made no sense to Daniel or to 99% of the other physicians with whom he discussed this issue. The FDA became the butt of jokes within the infectious diseases community. But companies and physicians could once again work on new antibiotics against the MRSA superbug that was causing 70% of these skin infections at that time in the US.

Next, the FDA tackled clinical trials of antibiotics for the treatment of pneumonia. Pneumonia is divided into two types – that beginning in the community in patients living at home and that beginning in a health care facility like a hospital or a nursing home. This resulted in another five years of fiascos one after the other.

In 2008, the FDA convened a two-day workshop on community-acquired pneumonia and how clinical trials could be designed to show that antibiotics work for these infections. To Daniel and almost every other infectious disease physician, the question was nonsense to start with. In the days prior to antibiotics about 1 in 5 to 1 in 3 patients with pneumonia would die of their infection. Mortality was worse among older patients and among those where the bacteria had also invaded the bloodstream. In these latter cases, up to 4 of 5 patients would die without antibiotics. Antibiotics changed all this. Today, only 2 out of every thousand patients treated for pneumonia will die. To Daniel and everyone else (almost) this seemed like prima facie evidence that antibiotics worked.

But the FDA wanted to know how well they worked in modern times. They came up with three suggestions. First, they wanted antibiotic trials to focus on more severely ill patients. Daniel actually agreed with them on that one. But then the FDA actually proposed that placebo controlled trials be carried out in pneumonia. That would be like taking a gun with 5 bullets loaded into six chambers and trying to play Russian roulette. Daniel and others likened this to the idea of carrying out a placebo-controlled trial of parachutes to demonstrate that parachutes save lives. Everyone already knew that, taking all comers, without antibiotics the pneumonia could be expected to kill up to 1 of every three patients. The losses would be higher if you restricted the study to those with more serious illness as the FDA was also proposing. This was clearly an unethical approach and everyone except the FDA was convinced that this was true.

Then the FDA wanted to change the endpoint, which had previously been cure at a follow-up visit after the end of therapy, to either mortality or to an earlier endpoint like improvement of symptoms within the first few days of treatment. They chose the latter because when they again looked at trials conducted in the 1930s for sulfonamide antibiotics in patients with pneumonia, the greatest difference between treatment and the sugar pills occurred during the first few days. In fact, the effects of sulfonamides were dramatic. About 85% of patients who received no antibiotic still had fever on day three, but only 2% of those that received antibiotics still had fever – this is a treatment effect of 83% (=85-2). This effect remained very large for over a week of treatment duration.

The rule is that the so-called non-inferiority margin that determines the size of clinical trials, cannot ever be larger than one-half the treatment effect. That means that for pneumonia, you could easily justify a margin of 20% - one quarter of the treatment effect. The FDA always argued that physicians would never accept allowing a trial where the new drug could be up to 20% worse than the old antibiotic. But, Daniel argued – "that statement is nonsense. Its not that the drug would be 20% worse, its that the statistical chance that it would be 20% worse would be greater than 5%. That's not exactly the same thing." But the FDA was not listening for a change. The margin would always be 10% increasing trial size by three-fold.

Next, the FDA wanted to limit the trials to patients who had a bacterial caused of pneumonia documented by growth of the bacteria from a sample from the patients' lungs. Daniel and his colleagues in the infectious disease society as well as those in industry were so taken aback by these proposals that they could barely speak. Luckily, the FDA had released an outline of their proposals several days prior to the meeting so Daniel and his colleagues could prepare their responses.

The Infectious Diseases Society quickly did away with the idea of a placebo controlled trial by showing data from the pre-antibiotic era and coming to the unanimous conclusion that they would refuse to participate in such a trial for ethical reasons. Daniel pointed out that with the mortality rates from pneumonia being so low in modern times, it would take between 50,000 and 200,000 patients to compare two antibiotics in terms of mortality as an endpoint. Since previous trials enrolled something like 800 patients each and still cost around $30 million, it was highly unlikely that a trial where mortality was the endpoint would ever be undertaken by anyone.

The FDA then stated adamantly that only patients with bacterial pneumonia proven by growing bacteria from the patient could be studied. They would not move even after the infectious disease society pointed out that in the US Medicare population, only 7% of all patients had culture proven disease. A review of recent trials submitted to the FDA showed that less than a third of patients studied had culture proven disease. Therefore, to have all patients in this category, you would have to multiply the trial numbers by at least three fold. Again, these numbers would render trials infeasible because of cost and because of the length of time it would take to complete them.

Another requirement the FDA was adamant about was one regarding the use of prior antibiotics. The problem is a complicated one. According to the FDA, in at least one earlier trial where this was examined carefully, the use of prior antibiotics altered the results of the trial suggesting that a new antibiotic worked for pneumonia when other evidence from the same trial indicated that was not true. The problem is complicated and remains a little controversial. Many patients, when they fall ill, call their physicians. Sometimes, the physician will prescribe antibiotic pills to be started at home. Then the patient worsens over the next several hours and ends up in the hospital emergency room. If they have pneumonia, they could be entered into a trial of a new antibiotic. But if the few pills they might have taken would alter the outcome suggesting that the new antibiotic worked when it really didn't, clearly this would be a bad idea.

The other problem is that these days, to enter into a clinical trial you have to sign a consent form. Of course, this is a good thing. But the forms these days are long – 15-20 pages. Plus, the patient frequently has to wait for various tests to come back from the laboratory to be sure that they meet all the inclusion criteria for the trial and that they don't have any of the exclusion criteria. This whole process takes 6-8 hours. But it has been shown that for seriously ill patients with pneumonia, delays beyond just a few hours increase mortality rates for the disease. Emergency room physicians are required to treat these patients quickly for this reason. Therefore, most patients get at least one dose of another antibiotic before they can be enrolled in a trial anyway. Daniel tried to point all this out to the FDA – but to no avail. This requirement would simply exclude most patients from any clinical trial of an antibiotic in pneumonia.

At the end of the two-day workshop, there was no clear pathway available for companies to study antibiotics for pneumonia. The FDA would hold additional meetings on this topic over the next several years. Each one would show a little progress in terms of the FDA backing off some of its unreasonable demands. It would take another five years for the FDA to come back to something feasible.

Two companies, Advanced Medical Sciences and Replidyne, had agreed to clinical trial designs to study community-acquired pneumonia with the FDA before the Ketek scandal in 2006. By the time both companies completed trials in 2008, the FDA had reversed course and informed both companies that the trials they had carried out would no longer meet the FDA's requirements for approval. Both companies declared bankruptcy since they were unable to raise sufficient funds to carry out an entirely new set of trials. This move by the FDA, as much as anything else, sent the industry scrambling for cover. If you could no longer trust the FDA to stick to their given word, how could you possibly invest precious research dollars in antibiotic development? The FDA could pull the rug out from under you at any time.

Hospital-acquired pneumonia was tackled next. Here the situation was even more difficult than any other yet faced by the FDA, by industry and by infectious disease clinicians. Daniel and his colleagues had been testing new antibiotics for these infections for years. But as the hospital patient population became more and more complicated with more and more underlying diseases, physicians had an increasingly difficult time deciding how to even diagnose the disease. It used to be that the presence of fever or other signs of systemic inflammation plus a new shadow on a chest X-ray would be enough to say that someone had a pneumonia. But now, patients with chronic heart failure had fleeting lung shadows all the time without infection. Those with other severe underlying problems frequently had both signs of inflammation and lung shadows without clear pneumonia when examined more carefully. But Daniel had always thought that these signs plus symptoms of pneumonia like cough or more importantly the production of respiratory secretions full of bacteria should be enough – and he hadn't changed his mind. There is no reason for well, normal patients to have large numbers of bacteria in their lungs. Daniel felt that this constellation should be enough to establish the diagnosis. But many lung specialists and intensive care doctors disagreed with Daniel's approach and the FDA was caught in the middle.

The other problem the FDA had was how to define an endpoint for treatment of this disease. Again, these were very sick patients who were frequently on artificial ventilation and often had serious underlying disease as well as their pneumonia. They often died. The FDA wanted to make companies prove that they could improve mortality rates by treating the pneumonia.

At one key advisory committee meeting, the FDA statisticians presented their preferred approach to a mortality endpoint. They wanted to use a statistical method called an odds ratio method. They asked that only patients so sick that 20% of the population would be expected to die be studied and that companies show that they could decrease this mortality by one-third. Daniel did not understand the statistics, but he did understand the problems with such a trial. First of all, patients this sick would have all sorts of complications some of which would be blamed on the drug being studied. But if a company wanted to study less sick patients, with this particular statistical method, the trial population went up enormously. Even following the FDA's advice, you would end up studying thousands of patients. Daniel knew that the largest trial ever conducted in hospital acquired pneumonia studied 1500 patients in over 300 different hospitals around the world, cost almost $200 million dollars and took almost two years. The trial the FDA was demanding probably could not be done at all, would require much more time, many more patients and would cost even more money. No company in the world would ever undertake such a trial.

The other major problem was that mortality was not a clean endpoint in Daniel's mind. Daniel knew of several studies that looked at mortality in patients with pneumonia acquired in the hospital. They all indicated that roughly half of the deaths could be attributed to causes other than the infection being treated. So in a trial comparing two antibiotics, if the endpoint were mortality, half of the patients would fail for a reason other than the antibiotic. This would actually end up making the two drugs look more similar and defeat the whole point of the trial. Daniel tried to make this point on several occasions both privately to the FDA and publicly at their advisory board meetings. But no one was home.

The one type of infection that was most important to physicians and patients would go without any new antibiotic for the foreseeable future because the FDA was demanding a completely infeasible trial that no one would ever carry out.

Daniel argued that instead of mortality, the FDA could simply go back to the way such trials were done in prior years where clinical response to treatment was how a new antibiotic was measured. There were lots of recent data showing that this could be done and these data came from modern clinical trials. But the FDA, listening to its statisticians and the intensive care specialists, wanted mortality as an endpoint.

Daniel wrote articles exposing the insanity of the FDA's approach and proposing what he thought were common-sense alternatives. He tried to make just a few key points beyond the specifics of clinical trial design. Daniel took pains to point out that not having new antibiotics to treat infections resistant to current antibiotics was, in itself, a serious safety issue. By blocking antibiotic development, the FDA was threatening the safety of future generations of Americans. He emphasized the infeasibility of current FDA requirements for trials in pneumonia in particular.

In 2011, Daniel was invited to give a lecture at a small meeting of FDA and academic experts that takes place every year at the University of Texas in Austin. The title of his lecture was "FDA – on the road to global irrelevance." In this lecture, Daniel noted that the US market for antibiotics was falling while markets in Asia were climbing. The overall market was flat. The reason for the market stagnation in the US was a lack of new antibiotics. There were no new antibiotics because there were no longer any companies developing them and this was partly because there would be no way of getting FDA approval to market them in the US anyway. One of Daniel's colleagues said that in the future, new and needed antibiotics would be available to the Chinese in Beijing but not to Americans in Washington. He also noted that continuing down this path, the antibiotic regulators would soon have nothing to do. In fact he had been hearing from FDA staffers that, even though they were getting some work around early clinical trials, very few sponsors were developing drugs to the point of actual market approval. Daniel suggested at this meeting that the FDA needed to hit the reset button on their approach to antibiotic development. Unfortunately, the FDA staff who attend the meeting were all up in their rooms during Daniel's lecture. He never knew whether they got his message or not.

Daniel repeated this message at a meeting on emerging infections in Atlanta the same year. At that meeting, he actually ran into the FDA commissioner and was able to speak to her about his ideas. She wasn't too happy with his critiques – or at least that is what Daniel thought. Daniel did, however, keep in touch with the FDA. He had maintained a good, respectful, working relationship with the antibiotics team at FDA. He had regular telephone conferences with them and saw them at meetings. They even used him to float ideas out to antibiotic developers in industry to see if they would be feasible if incorporated into trial designs.

Later, Daniel was invited to attend a meeting at the Pew Charitable Trust. The Trust was lobbying for legislation that would improve the situation for antibiotic development in the US – or at least that is what they thought. That legislation was called the Generating Antibiotic Incentives Now or GAIN Act. This law, which ultimately passed, had two major components. One was a so-called monetary incentive for companies to develop new antibiotics. Here, if such an antibiotic was deemed to be needed to fight resistant infection, it could have an additional period of market exclusivity (protection from generics) in the US. The problem with this for Daniel and for the industry was that if your exclusivity already lasted 15 years, by the time you corrected for inflation, the last five years of additional exclusivity were worth little in today's dollars. It would not be an incentive.

But the other part of the GAIN Act mandated the FDA to provide for feasible pathways for the development of these drugs and to provide a mechanism for their rapid review and approval. Daniel thought that this portion of the bill was the only thing that made sense. But the bill was not very specific and left much up to the agency. The FDA had still not shown that it was capable of actually carrying out such policy changes by the time GAIN passed congress. At the Pew meeting, Daniel noted the lack of monetary incentive in the bill, but emphasized the potential importance of having the FDA provide feasible pathways for antibiotic development.

In 2012, Daniel was invited to organize a symposium at the Gordon Research Conference on New Antibiotics. This was the same meeting where, five years earlier, the FDA had announced that they no longer understood how to carry out trials for antibiotics in the treatment of serious skin infections. Daniel invited the director of antibiotics from the FDA. This was the same person that at the previous Gordon Conference announced that the FDA did not know how to design trials for skin infections. Daniel also invited the head of antibiotics for the European regulatory agency and a prominent head of antibiotic development in industry to be the other speakers. Daniel moderated the discussion. Europe was clearly way ahead of the FDA and the FDA director knew it as did everyone in the audience.

The FDA director's presentations at the symposium made it seem like the FDA wanted to change and make trials both reasonable and feasible. It was clear, nevertheless, that they were going to stick to their crazy notion of using pre-antibiotic era data to justify their designs. Daniel thought that this would only lead to more conflict – but at the same time, he thought he was seeing a small light at the end of a dark tunnel. The speaker for industry was John Rex from Astra Zeneca. He presented a plan for development of antibiotics that provided for a tiered approach going from standard trial design through a very accelerated and more risky plan to get new antibiotics for resistant infections to the patients who needed this treatment very quickly and at lower cost. There was almost universal support for this carefully thought out approach. Even the FDA and the European regulators seemed to be behind the plan outlined by John.

At the conference, Daniel and the FDA director frequently took their meals together. They discussed the FDA's current approach and Daniel's ideas for change. The director told Daniel that he would soon be invited to participate on a task force being set up by the FDA to re-examine antibiotic development. Daniel was surprised both by this sudden openness and by what he discerned was a changing view at FDA. He wondered what was going on politically within FDA.

Two weeks later, Daniel received a note from the Brookings Institute inviting him to participate on a task force on antibiotic development. The task force was being set up at the behest of the FDA and the FDA had requested that Daniel be invited to participate. The first meeting took place in May, 2012, about six weeks after the Gordon Conference. The first speaker was Janet Woodcock, the Director of the Center for Drug Evaluation and Research in the office of the Commissioner for FDA. Her first sentence after thanking everyone for coming was to announce that the FDA was going to "reboot" their antibiotic development program to focus on patients with resistant infections and to provide feasible pathways for the development of antibiotics. She admitted that the FDA's approach to antibiotic development since Ketek had been a failure and said that the agency realized that they were going to have to do things in a completely different way.

Daniel and the rest of the task force members sat open-mouthed, slack-jawed and many were secretly jumping for joy. Daniel and several others at the meeting pointed out that the FDA would also have to work on more traditional drug development pathways to allow for new antibiotics to come to market via those pathways as well. Janet seemed to agree that that would be necessary. The rest of the meeting was dedicated to exploring in greater detail how the development plans outlined by John Rex and his colleagues could be translated into feasible trial designs. But Daniel felt that Dr. Woodcock's speech would be a historic turning point for antibiotic development at the FDA and he hoped that this would ultimately translate into a greater enthusiasm for investment and for research within the pharmaceutical industry.

Within the next 18 months, both the European regulators and the FDA would release new guidance on developing antibiotics targeting seriously ill patients with resistant infections. The trial designs outlined in the guidance documents would call for small trials that could be performed quickly. The agencies acknowledged that these trials, if successful, would increase the safety risk of drugs approved through this pathway. The documents incorporated much of what Daniel, John Rex and other colleagues had been discussing for the previous decade.

While Europe had always maintained a feasible traditional pathway for antibiotic development and clinically relevant endpoints like cure of infection, the FDA with its infeasible trial design requirements fell far behind the older continent. The FDA would still have to take on the traditional development pathway for antibiotics to truly reboot. Daniel, along with the infectious diseases society, the pharmaceutical industry and many others, focused his efforts with the FDA at getting this accomplished. By 2014, even the traditional pathway for antibiotic development was being rebooted. Even though the FDA remains stuck with less relevant endpoints, the trials now proposed for most traditional infectious diseases like urinary tract infection, pneumonia and skin infection are all feasible.

Chapter 3

Money Talks

With the easing of regulatory restrictions on antibiotic development in the US, Daniel was hoping for a surge of company interest in antibiotics. In fact, during the years of the FDA reboot process, two companies that had previously abandoned antibiotic research announced that they were restarting their efforts – Sanofi-Aventis (the Ketek victim) and Roche – the first company to announce that they were renouncing antibiotic research back in 1999. But other companies still dithered. Pullman and Astrazeneca both considered pulling out.

In thinking further about this, Daniel realized that the second half of the equation had still not been addressed – money. The key for companies is return on their research investment for their shareholders. One of the several reasons cited by companies when they pull out of antibiotic research is that they are unable to make a sufficient return on their research investment in antibiotics. The reasons for this include trials that are just too expensive (now being resolved), the fact that antibiotics are taken only for a few days to two weeks compared to other drugs that have to be taken chronically, and the fact that many antibiotics are already generic and cheap – so it is difficult to charge a price high enough for a new antibiotic to adequately increase profit margins.

In recent years, a number of academic studies focused on this question of return on investment in antibiotic research. They all identified two mechanisms to influence the equation. The first was to provide funding to help reduce the company's costs of research. They called this a "push" incentive. By reducing the investment, it would be easier to get a return in the marketplace since the hurdle to a positive return would be lowered. The other approach would be either to pay higher prices; a so-called "pull" incentive. One pull incentive frequently discussed is a guaranteed government purchase at the time of approval of a new antibiotic active against resistant infections.

Daniel had been shocked on several occasions by the prices being demanded for new drugs to treat cancer. These drugs almost never led to cures of the disease and frequently prolonged life by weeks to months, yet they were priced at $50,000-100,00 per course of therapy. When Daniel thought about antibiotics, he realized that they sold for pennies and actually cured disease within a few days. There was clearly something wrong with this picture.

Based on these considerations, Daniel got together with some marketing colleagues to examine this problem more closely. They invented a hypothetical new antibiotic that would be active against only one species of bacteria – similar to the one Daniel ran into at Pepcore. Research using databases in Europe and the US demonstrated the number of patients that could potentially be treated with such an antibiotic. Daniel and his collaborators then were able to calculate a price that would be required to provide an adequate return on investment for such a product. It was in the range of $15-30,000 per course of therapy – still cheaper than many cancer drugs, but unheard of for antibiotics.

The group published these findings. Shortly after their paper was published, Daniel got a phone call from Doris Ellsman. Doris was a tall, thin woman with short reddish blond hair and a Liverpool accent. She was a highly respected microbiologist and pubic health authorities in Britain and around the world took her opinion very seriously even though she was not a physician.

"Hello. Daniel Simon here."

"What the hell are you talking about?"

"Doris? Is that you? What are YOU talking about?"

"Your paper you bloody idiot. No one will ever pay those kinds of prices for an antibiotic."

"Oh? And what would you have them do to treat these infections resistant to virtually every antibiotic available?"

"Rather than pay those prices, they should be treated with a combination of antibiotics including colistin."

"Colistin?" Daniel was incredulous. He knew that Doris frequently held strong feelings and this kind of conversation was nothing new for Doris or for Daniel. "Doris – first of all there is precious little data that suggests that colistin will even work for these infections even though it has been around since the 1970s. I used it myself when I was training back in Cleveland and virtually all my patients died of their infection with high levels of colistin in their bloodstream. Not only that, but 35-50% of patients treated with the recommended doses will suffer kidney damage. So what are you suggesting?"

"Well, we just aren't going to be paying those prices – that's all."

"OK. So you would rather pay the societal price of preventable death and take care of survivors on kidney dialysis machines for the rest of their lives. Is that what you would prefer? There is no free lunch, Doris."

With that, they politely agreed to disagree on this point. The conversation, though, reminded Daniel that convincing payers to provide the kind of price he was contemplating on a global scale might not be straightforward.

The Pew Charitable Trust also picked up Daniel's paper. They were lobbying the US congress to pass legislation directing the FDA to provide a specific pathway for the approval of such antibiotics. The Pew arranged a conference specifically to examine how such antibiotics could be priced by inviting payers and hospital pharmacists to a meeting in Washington. Daniel was a rapporteur at the meeting. The payers and others were presented the hypothetical antibiotic invented by Daniel and colleagues and were asked if they would support such pricing.

The insurers all indicated that they would agree to these prices if the advantages of the new therapy could be demonstrated. The data that they would require to provide such a price remained vague. Daniel assumed that the data that could be produced by the clinical trial of the antibiotic plus other information about the usual course of disease in patients treated inadequately would provide the kind of information on cost and benefit that the payers usually required.

Pharmacy managers were a different group, though. Within a hospital or hospital system, they had a fixed budget. Agreeing to bring on another high priced product might put them over-budget. This means that hospital or system managers – the bosses – would have to approve these extra expenses. With the agreement of insurers and Medicare to cover such therapy, though, the question should have been answered. But the pharmacy managers remained skeptical knowing how their own managers usually react to new, high-priced drugs.

Not everyone heard the same conclusion from the meeting Daniel found out later. Some heard the complaints of the pharmacy managers while others heard the acceptance of the insurers and Medicare. Once again, Daniel wondered how payers outside the US would react since he still had Doris' words clanging around in his head.

With a global picture in mind, Daniel began to call colleagues at various companies to see if anyone could explain to him how cancer drugs were able to garner such high prices around the world. He was never able to understand drug pricing in many countries. India was a particular challenge since they had so many generic versions of patented products, large numbers of simply counterfeit drugs as well as genuine branded products on their market. Plus, in India, you could often get antibiotics at a local pharmacy without a prescription. How could one manage pricing in an environment like that?

The question of pricing and reimbursement for antibiotics was taken up by a number of large companies and various non-profit organizations – especially in Europe. The Pew in the US continued to examine the issue of reimbursement for antibiotics as well. Daniel could see a guaranteed government purchase occurring in European countries where each country negotiated a single drug price nationally in any case. He did not see this occurring in the US where drug pricing is more like the Wild West. Although Daniel could not see how things would ultimately sort out, he was confident that the economic question around the return on the research investment for antibiotics would be resolved one way or another.

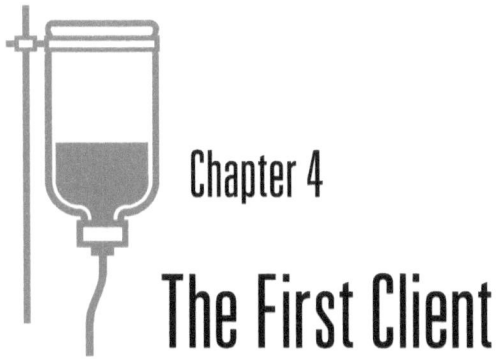

Chapter 4

The First Client

Daniel arrived in London to meet Deborah Cranston-Smith and her team in September of 2005. There were a number of players around the conference room table when Daniel arrived at Mitsubishi Ventures near the Strand in London. Most were Deborah's employees. Then there was Gerhardt Unger. Daniel had never met Gerhardt, but he had run a very successful biotech in Oxford, England called Calibodies. They specialized in finding and optimizing antibodies that could be used to treat various human diseases. Gerhardt knew a great deal about antibodies and something about running companies. Calibodies was sold to Gorman Ltd for over $1 billion several years earlier. Gerhardt was now the head of the British Biotech Association and on the boards of several companies. Calibodies had been an investment for Mitsubishi and its sale had made Mitsubishi and Deborah very rich. Gerhardt was obese beyond stocky, had strands of long, reddish hair that poorly covered an almost entirely bald head. His blue eyes could barely be seen behind thick lenses. He had a loud laugh and a very un–German, more British, understated humor. He also had a strong accent and Daniel occasionally missed what Gerhardt was saying and had to have things repeated too often for comfort.

Deborah herself presented the company that was planning to spin out a small antibiotics research group. Ullanger Pharma was one of the world's largest pharmaceutical companies and had its headquarters in Basel, Switzerland. The spin out would remain within the Ullanger campus in Basel for the first year or so until other laboratory space could be identified. The group had discovered two new antibiotics both of which were ready for their final toxicology tests and their final animal model studies before going into human clinical trials. They had ongoing research projects mostly focused on finding backup compounds for these two leads. Gerhardt then took over to discuss the two lead molecules. But Daniel found that there was so much missing information that he was unable to understand how these molecules could progress. He suggested that he meet directly with the team at Ullanger before providing any sort of opinion or guidance to Mitsubishi. Gerhardt looked irritated that his presentation was not satisfactory. The problem was that Gerhardt himself did not have a good understanding of antibiotics or how to develop them and therefore he had been unable to anticipate the kinds of information he would need to show to someone like Daniel.

The arrangements were made quickly and Daniel arrived in Basel a week later to meet with the Ullanger team along with Deborah and Gerhardt. Wolfgang Reiser, Hermann Hartz and Mona Harmon represented the Ullanger team. Wolfgang was tall, blond, thin, young and pretty. Hermann was a chemist and clearly the leader of the team. He was older, probably around 65 years of age, had dark straight hair that always fell over his eyes, and thick black framed glasses. Mona was quiet, had medium length dirty blond hair, and she was taller than everyone else in the room. The task of presenting the Ullanger compounds to Daniel fell to Wolfgang who was visibly uncomfortable. Daniel wasn't sure if Wolfgang's discomfort was because he had to give a key presentation in front of his boss and potential investors or if it was because he did not have confidence in his knowledge of antibiotic development or both. But it was clear to Daniel that Wolfgang was the least comfortable person in the room.

Wolfgang's presentation was more informative than Gerhardt's had been. But, in spite of having Daniel's questions ahead of time, answers to those questions were not forthcoming. Both Gerhardt and Wolfgang had shown that the compound, called UL-4215, was a member of a type of antibiotics used to treat infections in animals but that had never been developed for human use. This class of compounds had previously had issues with toxicity. But UL-4215 seemed to be well tolerated by the animals and therefore should be more safe than the similar drugs being used to treat animals. Carefully, Daniel repeated his questions from the previous week around the therapeutic dose for animals and initial toxicology showing how much drug the animals could tolerate. Wolfgang and the Ullanger team had not carried out the key animal model experiments to define the therapeutic dose, but Mona was able to show some early tox data to suggest that animals could tolerate a high dose of the drug orally. When Daniel inquired about intravenous therapy, Mona looked at the floor and quietly admitted that the drug was insoluble and could not be tested intravenously. It could only, therefore, be developed as a pill. Daniel said that this would not necessarily be a problem. Ullanger did have some data suggesting the compound would be active against key bacteria causing skin infection and pneumonia. But they had not tested it against a large number of bacteria from patients around the world, so they did not have a good idea as yet as to whether there was really any resistance already out there.

The second lead compound was related distantly to the penicillins and seemed like it would be safe based on early tox data shown by Mona. But it seemed to offer little advantage over other drugs already on the market as generics in terms of its activity against resistant bacteria. Again, the data on antibacterial activity was really only based on a few strains of each species of bacteria from Switzerland and there was no animal model testing of the compound. Daniel was skeptical about this particular compound, but the limited data available might not present a true picture of the drug's activity. For example, if the drug were very non-toxic, high doses could be used and bacteria that might otherwise seem resistant, could actually be susceptible under those circumstances. There was no way of knowing for sure without more data.

That evening, everyone went out to dinner together at Les Trois Rois in Basel overlooking the Rhine. It was a gorgeous evening with a beautiful sunset enjoyed with champagne and Rosti (Swiss grated and roasted potatoes). Daniel was not sure what exactly was being celebrated.

Deborah asked Daniel if he could provide them a summary of his thoughts on the two projects the next morning at Ullanger before catching his train back to Paris.

Daniel, the next morning, spoke about the lack of experience developing antibiotics within the Ullanger team and how such knowledge and experience would be required to take their lead compounds forward. He specifically noted the lack of robust data on the activity of the compounds against a global collection of bacteria with varied resistance properties and the lack of animal model data to help define the therapeutic dose that might be required in humans. Daniel did note that either compound might be an important addition to the armamentarium of physicians. He emphasized the potential that UL-4215 could provide an antibiotic in a pill form to treat infections with bacteria resistant to most other antibiotics. Since there was only one such drug marketed, Daniel thought this was their most important asset.

The team thanked Daniel for coming and for his efforts. Deborah said she would get back to Daniel shortly regarding follow-up to the meetings. She clearly expected Daniel to help with the further evaluation of Ullanger as a spin out opportunity.

The next meeting was back in the Mitsubishi offices in London. There was no one from Ullanger in the room. It was just Deborah, her team and Gerhardt. There was one new face – Clive Burns. Clive was a Brit from Manchester who Deborah had hired as a consultant on manufacturing. He had previously been head of scale up chemistry for Glaxo and later for Gorman. The goal of this meeting was to define the organizational structure of the new company and the number of employees that would be part of the spin out. There was also to be a discussion of which current Ullanger staff could fill which roles in the new company. The new company did not yet have a name – Deborah was awaiting the result of polling at Ullanger to find a name – so it was referred to as Newco. Gerhardt went to a white board and drew out his version of Newco's proposed organizational chart. Daniel shook his head. Gerhardt's chart had a CEO reporting to a "supervisory" board (the Swiss version of a Board of Directors), a Chief Financial Officer, a head of Business Development, and underneath he wrote biology and chemistry research. Daniel pointed out that they were within one year of submitting an IND and starting clinical trials on two new antibiotics. How will Newco deal with the regulatory agencies? Who will run the contracted toxicology studies? Who will design the early clinical trials? Gerhardt suggested that all this could be done by consultants. Clive was silent. Daniel asked him whether manufacturing could also be managed through consultants. Daniel felt strongly that in-house expertise in toxicology would be needed to supervise and later analyze critical studies done at outside contractors. He also argued that a regulatory affairs person be hired to help write and submit all the documents that would be required especially towards the last half of the first year of Newco's existence. He also suggested a full time project manager to organize the disparate tasks that would all have to fit together to get everything done in time. Finally, Daniel suggested that they hire a clinician not only to design trials, but to feedback to the scientists on what was most likely to be able to advance to the clinic and what data would be needed to allow that to occur. But on the organizational chart, for the most part, Gerhardt won the day. Mitsubishi clearly wanted to limit their investment if possible. Daniel thought they were being pennywise and pound foolish.

Daniel was pleasantly surprised, though, when Deborah agreed with his suggestion that they move sufficient staff from Ullanger to Newco that they could have a robust backup program for the two lead compounds in case either should fail. Such a team could also try to discover and bring new molecules forward so that Newco would be more than just a one trick pony. They agreed to move around 30 biologists and chemists to Newco including the current Ullanger toxicologist (in a bow to Daniel's org chart suggestion). Daniel would be tasked as a consultant to help lead their discovery, preclinical and clinical development efforts and he began to wonder what he had gotten himself into. All these considerations led to a budget plan over the first three years for the new company. This budget plan would guide the size of the initial investment in the spin out.

Given this plan, Daniel suggested he go back to Ullanger and start to get better acquainted with their scientists so he could understand how much expertise they had and therefore how much of Daniel's time this consulting job would require. Deborah agreed and Daniel was back in Basel within the month. This time he met with all the lab directors. Daniel discovered that they had very competent scientists who could carry out specific experiments, including a veterinarian who was running their in-house animal model work, but that none of them had even a fundamental understanding of the kinds of experiments that would be required to move an antibiotic into clinical trials. Even Mona, with her expertise in toxicology, had trouble seeing beyond her own narrow sphere of work so that she could place data she got back from her studies into the context of how and at what dose the antibiotic under study might benefit patients. The veterinarian was wonderful at doing animal model studies, but had little understanding of what models he should be using. The microbiologist could test bacteria for susceptibility to the antibiotics being discovered at Ullanger, but did not know how to proceed beyond the relatively few bacterial strains Ullanger had in house.

Based on these meetings, Daniel decided to embark on a series of lectures with slides to help the Ullanger scientists understand the tasks they would have to accomplish over the next year or so. His course in antibiotic development 101, the formation of an external advisory board of key scientists and opinion leaders, along with monthly meetings with the team would be the basis of Daniel's first year working with the Ullanger spin out.

It was November by the time all this was put in place. Deborah and her team were out trying to put together a consortium of investors where Mitsubishi would be the lead. And they had made good progress. Ullanger itself would make an investment and would have a seat on the supervisory board. The spin out was scheduled for January 2006. The name of the new company was to be Swisbiotics and Mitsubishi targeted an initial investment of about 30 million Swiss francs. Swisbiotics would also receive an additional 10 million Swiss francs from the Swiss government. The total initial investment would be around $50 million and change.

One of Daniel's first tasks with Swisbiotics was to have them contract with an academic lab doing the kinds of animal model studies that would allow them to calculate the therapeutic dose of UL-4215. Once they had these data, they would understand how to dose the drug – whether it had to be given multiple times per day or whether they would have the option of once daily dosing – and how much drug would have to be given. With this information, the data from the toxicology studies could be put in the context of treatment. Obviously, if the dose required to treat infections was greater than the toxic dose, the drug would have to be abandoned. He brought in an old friend, John Anders. He asked that John be a member of Swisbiotics nascent scientific advisory board and that he get the contract for these key animal studies. Swisbiotics did this quickly and John came over to Basel to give them a one half day course on how these experiments were designed, how dosing regimes were chosen and how the therapeutic dose could be calculated. The Swisbiotics scientists could not have been more enthusiastic learners and they quickly understood John's plan for UL-4215. The data came in several months later and showed that the number of doses given per day would not matter – it could be one, it could be four or more – the drug would be just as effective. He also showed that a dose of about 3-500 mg per day should be about right in people making assumptions about scaling from mouse to man.

When the toxicology data came in, Mona presented them to one of the team meetings when Daniel had joined by phone. Daniel could see her worried and anxious face over the phone line just based on the tone of her voice. She noted that at the mid dose and the high dose in rats, but not in dogs, in the key two-week study they would use as the basis for submitting for permission from the FDA to study the drug in humans, there were fatty deposits seen in the liver. Daniel asked – but was there evidence of inflammation of the liver or liver cell death? No. Could it be species specific and not relevant to humans? Yes. Daniel thought that this was probably not a so-called adverse event, that it could even be a species-specific problem and that these doses would be deemed safe by the regulatory agencies. Mona was not sure. Daniel suggested that they get an outside consultant to help them decide because if the mid-dose in rats was deemed toxic, it would be very hard to get UL-4215 tested in humans. So Swisbiotics hired a toxicology consultant in the US who would also act as the US sponsor for Swisbiotics' interactions with the FDA. You have to have such a US sponsor with a US address to get FDA approval to do studies in the US. The toxicology consultant agreed with Daniel. He helped write the final report from the toxicology study that would show that the proposed therapeutic dose in humans would be safe and that the highest dose studied in dogs and rats was safe. They based their proposed human starting dose for the first trials in people at one-twentieth the highest dose that was safe in animals.

Then the next piece of toxicology data came in. Swisbiotics knew from earlier studies that UL-4215 could bind to a receptor (HERG) in the heart muscle that was involved in conduction of the electrical signal in the heart. As part of the routine safety studies required before embarking on clinical trials, they examined the effect of UL-4215 on blood pressure, pulse and heart rhythm in monkeys. The highest dose in monkeys showed that the electrical signal seen on the electrocardiogram was slowed (prolonged QT interval). This was a red flag for regulators – but less so for physicians. This signal was, infrequently, associated with arrhythmias of the heart. The dose in monkeys that caused this effect was higher than the therapeutic dose in people, but only by about two fold.

Given all this, Daniel suggested that before actually submitting their Investigational New Drug application to the FDA, that Swisbiotics meet with them to be sure that the FDA agreed that it would be possible to proceed with developing UL-4215. This meeting occurred in mid-2007. The FDA agreed with Swisbiotics plans for UL-4215 and agreed that the risk of liver and cardiac toxicity was low and that the starting dose chosen by Swisbiotics, now 40 fold lower than the lowest toxic dose in animals, was appropriate. The IND was submitted by the end of 2007 and clinical trials started in early 2008.

After pushing Swisbiotics CEO, Gerhardt Unger, for the previous two years, he finally agreed to hire a clinician to help contract, design, run and interpret the clinical trials of UL-4215 that would start soon. He brought in several job candidates and Daniel was asked to interview them. Daniel found that the candidates available who were willing to work in Basel had little previous experience in running early phase clinical trials or had no experience in doing so with antibiotics. There was really only one candidate with prior antibiotic experience, but he had been more of a project manager than a clinician running trials. Daniel was so frustrated by this that he was ready to ask that the clinical development group be moved to the US where he knew they would be able to hire top quality people. But Swisbiotics was staying in Basel and hired the project manager to develop UL-4215. Obi Nkruma was born in Nigeria, trained in London, England and had worked most recently for Gorman in the US. He had been involved in managing late stage clinical trials of one of their antibiotics that subsequently succeeded in getting on the market.

The first trials started in early 2008 under the supervision of Swisbiotics new Chief Medical Officer, Obi Nkruma. The data came in in blinded fashion. Each cohort had eight volunteer subjects. Six would receive 4215 and two would receive placebo. When the data came in the placebo patients were mixed in with those receiving the drug. In this way, there was no bias in the reporting of adverse events. Blood chemistry results came in pretty much at the same time as dosing completed. The data on blood levels of drug in the volunteers came in a little later so that at the end for each cohort, the Swisbiotics staff knew who had received drug and who had not. The physicians caring for the volunteers on the study unit did not know who was who until the entire trial was completed. After the first two cohorts getting increasing doses of 4215, it was clear that there was going to be a problem. The drug was being given simply as drug powder poured into a capsule. Given this way, there was already considerable variability in the drug levels seen from one subject to the next. The study was designed so that once a dose of drug had successfully been given as a single dose, a 10-day multiple dose study at that dose could then start. This staggered sort of design saved about a month in the entire plan and had been strongly suggested by Daniel.

The fourth cohort would be the first to see a dose that would probably be therapeutic. In the multiple dose study of this dose, around 300 mg per day, there was tremendous variability. Some subjects had almost no drug in their blood, some had therapeutic levels and others had two fold more drug than expected. There were no side effects in those receiving drug compared to those getting sugar in their capsules and, specifically, there was no problem with liver or heart toxicity that were being closely monitored by the Swisbiotics team. But the variability in the amount of drug absorbed was a serious problem. At a given dose, some patients would not get therapy at all and others might be getting toxic levels of drug in their systems. Daniel suggested that Obi call a halt to the trial and that the team rethink their approach. Both Obi and Gerhardt were reluctant to stop. Obi thought that perhaps the variability would iron itself out with higher doses. Gerhardt did not want to have to report a clinical hold on his lead drug candidate to his board. Daniel argued – "What if the variability does not improve? What if we get some patients with toxicity from unpredictably high exposures? What then? Lets stop now."

Gerhardt and Obi were not convinced. They suggested that their scientific advisory board be convened to review the data. Daniel, as chairman of the Board, arranged a meeting – but it would have to be several months away as the Board consisted of very well known and very busy physicians and scientists. Obi would convene a total of two advisory board meetings to review data coming from this one early stage trial in volunteers. That process would eat up the better part of a year. Daniel was not willing to wait.

The Swisbiotics team convened in Basel – Daniel traveled over for the meeting. They reviewed the data. Daniel asked Clive Burns, Swisbiotics' manufacturing consultant, to come to the meeting. Daniel felt that their only option would be to find a new formulation of the drug that would avoid this variability and provide for smooth and reliable absorption of the compound from the intestinal tract. Daniel had pushed to have this work started long before Swisbiotics started their trials, but they had balked for budget reasons. Once again, thought Daniel, pennywise and pound foolish. Clive agreed with Daniel and agreed to work on developing a new formulation. He said it would take about six months before they would know whether this was even possible.

At the same meeting, Daniel inquired as to where the research team was on finding a backup for 4215. The team had been working on this and had a list of properties they wanted in a backup. First, they wanted something that could be given both intravenously and orally. They wanted something that would not have the risk or at least would have a lower risk of heart toxicity. They wanted a drug where the pill form would not have the variability seen with 4215. And they wanted something that was more active against resistant strains that 4215 was. The team reported that they had been carrying out a search of older compounds from their library and at the same time had been synthesizing new compounds. One of these new compounds looked like it would fit the bill. They called it SW-1212 – SW for Swisbiotics and 1212 for the date it was synthesized. It was much more soluble in water than 4215, did not bind well to the heart receptor, HERG, and was absorbed from the intestine in rats. Given intravenously, the drug was well tolerated by the rats and the same seemed to be true when the drug was given orally. Daniel was so pessimistic about 4215 that he wanted the team to get 1212 on track for a new IND within the next 9 months and into clinical trials by mid 2009 in case 4215 could not be formulated successfully. Even if 4215 could be formulated, 1212 looked like such a superior drug, that Daniel wanted Swisbiotics to pursue it in place of 4215 assuming the animal models of infection and the toxicology studies went well. The team was in agreement. The only objections seemed to be coming from Obi who was spinning his wheels obsessing over the ongoing trial data. Even Gerhardt was convinced to go along with Daniel's plan.

At this point, Wolfgang Reiser came to Daniel with a problem. Wolfgang had been given the title of head of discovery research at the time of the spin out. During this time, he had been reading voraciously and bombarding Daniel with questions on antibiotic discovery and development. His problem at this point in time was Obi. He was angry. Daniel would learn that Wolfgang could have a short fuse – something he had never personally experienced but that he had heard about during informal conversations with scientists and even Gerhardt at Swisbiotics. According to Wolfgang, Obi was unable to clearly supervise the clinical trial, frequently did not come to work and was dumping much of the load on Wolfgang, who, like Daniel, was a physician and a scientist. Daniel thought about Wolfgang's complaint briefly and realized that he must be correct. Obi, since the beginning of the trials, had made absolutely no progress. There was no clinical development plan for 4215 in the case that they could solve its variable absorption problem. What trial would be next? What was needed to prepare for that trial. Who would carry it out? Where would it be performed? None of these questions were even being considered even though Daniel had repeatedly raised them at team meetings. Daniel even had some feedback from his advisory board members asking what was going on with planning beyond the volunteer trial. Daniel encouraged Wolfgang to discuss his concerns with Gerhardt and, if that yielded no results, to go to the supervisory board. The nature of the supervisory board was such that the key high ranking managers, including the CEO, Gerhardt, the Chief Financial Officer, the Head of Discovery and the Chief Medical Officer were all members of the supervisory board. So any of them could suggest agenda items and make presentations to the board. Usually, proper etiquette required that all had to be cleared by Gerhardt first – but if one felt strongly, a manager could go over Gerhardt's head directly to the Board.

The first inkling that Daniel had that all was not right at Swisbiotics was that Wolfgang was unable to move Gerhardt to do anything about the Obi problem. Wolfgang, as he had warned Gerhardt, brought the subject up to the supervisory board who, in turn, agreed with Wolfgang's analysis of the situation and instructed Gerhardt to speak to Obi. Gerhardt did so, but apparently the conversation was either so bland as to be ineffective or Obi simply ignored his instructions to develop a clinical plan for 4215 since one never materialized.

Obi was amicably separated from Swisbiotics three months later. Things in Europe were not the same as in the US. There frequently were strict requirements around notice periods. To separate someone (fire them) you were going to be obligated to provide them with a minimum of three months and frequently six months notice and salary plus vacation time owed. And in Europe, vacations were considerably longer than in the US. Daniel never heard how much this amicable separation cost the company. The relationship between Wolfgang and Gerhardt deteriorated in a way obvious to everyone in the company as a result of this clash.

Swisbiotics was left with an unfinished volunteer trial of 4215, a race to finish preclinical work on 1212 and no one to do any clinical planning. They embarked once again on a search for a chief medical officer who would be willing to live in Basel. This time they struck paydirt if not gold. They located a physician, David King, who had been involved in running early stage clinical trials in a variety of therapeutic areas including antibiotics for Astra-Zeneca, and later for a small biotech in Boston where he lived with his wife and son. The small biotech where he was working had folded as their only product had failed in clinical trials. David was willing to move his wife and son to Basel. Swisbiotics jumped on this opportunity and hired him. Daniel had interviewed him and quickly understood that he was not an expert in antibiotic development, but he did have a good command of early stage clinical development and clinical trials in general. Daniel felt that David could learn the subtleties of antibiotics without much of a problem and supported the decision to hire him. Another factor playing in Daniel's calculation was that Daniel himself was unable to take on much more work as his consulting business had become extremely demanding.

After David joined, he and Daniel spent a good deal of time together working on a clinical development plan that could fit, with a few modifications here and there, either 4215 as an oral only drug or 1212 as an IV and oral drug. The plan called for a first trial in patients (as opposed to the volunteer trials on hold at this point) with severe skin infections. The patients would have to be sick enough to require hospitalization. The study would be conducted in the US because there was a large population of patients infected with the MRSA superbug who could be included in the trial and demonstrate that either 4215 or 1212 could be used to treat these difficult infections.

Even though David got this job done, Wolfgang had concerns about David's work ethic. David, like Obi before him, spent a good deal of time out of the office. David, though, unlike Obi, seemed to be either working from home or visiting potential study sites – so Daniel was not sure that he shared Wolfgang's concern.

By early 2009, enough data had been collected on SW-1212 to file an IND with the FDA. John Anders again carried out the animal model testing. His data suggested that like 4215, the drug could be dosed either once or multiple times per day. The regime would not effect how well it worked as an antibiotic. He also showed that the dose could be less than one half the dose they originally contemplated for 4215. The toxicology studies demonstrated that the IV drug was very well tolerated and there was no fat accumulation in the liver as had been seen with 4215. Data also showed that 1212 had a lower affinity for the HERG heart receptor and that, in monkeys, although there still was a signal of decreased conduction in the heart, it occurred at much higher doses than seen for 4215. 1212 was also able to be absorbed from the intestinal tract of mice, rats and monkeys suggesting that a pill form would also work for people. Taking all this information into consideration, John Anders, David King, Wolfgang, Daniel and the rest of the team formulated a plan that allowed for rapid testing of the drug in volunteers followed within 6-8 months by an actual trial in patients. All these data and the Swisbiotics plans for 1212 were submitted to the FDA in early 2009 and trials in volunteers were planned for February. At the same time, attempts were being made to make a new pill for 4215 that would avoid the problem of variability. These efforts failed entirely. This failure plus the other concerns around 4215 was the final nail in its coffin. UL-4215 would be laid to rest and SW-1212 would take over.

Swisbiotics was not a one trick pony as the investors had originally foreseen. The discovery and rapid preclinical development of the backup for 4215 demonstrated how right they were to fund enough scientists to carry out the backup program. At the same time, Swisbiotics pursued two other programs. One involved a cephalosporin antibiotic, similar to the penicillins. It was licensed by a small pharmaceutical company in the US who took it into trials in volunteers. In these trials, 30% of subjects receiving a therapeutic dose of the drug developed a skin rash and that was that.

The third program involved a drug closely related to 4215 and 1212, but that would be used topically. Swisbiotics thought initially that a topical antibiotic would be a strong product because a similar product was the best selling such topical antibiotic on the market with $300 million in global sales. They asked Daniel his opinion on how the drug should be studied to bring it to market. Daniel thought that a quick way to a lucrative market would be to show that the drug prevented MRSA superbug infections when used to eliminate the MRSA carrier state from asymptomatic patients who were going to have surgery. If such patients developed a post-operative infection it was almost always caused by whatever strain of staph they carried – MRSA superbug or not. If applying Swisbiotics' topical to the nose (where staph is carried) could eliminate their carrier state, they should not get an infection or at least they should have a much lower incidence of infection than those not treated. The best selling product was the only one approved for this use.

Daniel suggested that they consult with his old friend Roger Butler to see if the market was worth it. In fact, Roger quickly did a study for them showing that the treatment to eliminate the carrier state accounted for less than 5% of sales for the marketed drug and that Daniel was wrong. Such an approach, although possibly allowing a quick entry to the market, would never bring them a return on their investment. Because Swisbiotics felt that this program was less valuable than their program on systemic antibiotics, and because their budget was limited, it received a lower priority and less funding. But over the years there had been a few potentially interested partners. All, ultimately, declined – so the topical languished. This meant that, going into 2009, Swisbiotics' most valuable asset would be 1212.

This time the IV form of 1212 sailed through its first trials. At very high doses, the effect of 1212 on heart conduction could be seen in the volunteers. Knowing that this was a possibility, David had designed the trial, at the insistence of Daniel, to include very careful, triplicate electrocardiograms to monitor heart conduction during dosing. They obtained blood levels of the drug at the same time as the EKGs and could relate exposure to the drug to any effect on the heart. The heart conduction time problem was related to the concentration of drug at any one time. So a single high dose given as a quick bolus injection was most likely to cause the problem. David designed the trial so that the drug would be given over a one-hour infusion intravenously and the dose would be divided into two daily doses. The data from this trial were so convincing that the FDA and eventually the European regulators agreed that no further systematic testing for heart toxicity would be required of Swisbiotics for SW-1212. To have this in writing from the regulatory bodies was unheard of at the time and made a very positive impression when it came to discussing 1212 with potential large pharma partners.

The only problem with 1212 was that some of the subjects had pain with the IV infusion and a few had phlebitis – actual inflammation of the vein where the drug was administered. To alleviate this, Swisbiotics, under David's guidance, undertook to change the IV formulation and, while awaiting a new formulation, to simply dilute the drug further and administer it over a longer period of time. All this worked well – problem resolved. Swisbiotics was ready to enter a clinical trial on patients with severe skin infections by the end of 2009. The trial started in December.

SW-1212 was more potent against bacteria, including the superbug MRSA, than most other antibiotics on the market or in development. This was confirmed in large studies of well over 1000 bacterial strains from all over the world. Swisbiotics and Daniel were sure that 1212 would work in patients with severe MRSA skin infections.

The trial was randomized and double blinded. No one would know which patient was getting which drug except a study pharmacist at each study site. This was required in case there was a problem and the code had to be broken on an emergent basis. There was a comparator drug, vancomycin, that many considered the gold standard because it was the most widely used antibiotic to treat MRSA infection. Part of the reason it was so widely used was that it was generic and cheap. Daniel argued strongly for using linezolid or Zyvox as the comparator. Zyvox came in both an IV and a pill formulation just like 1212. Zyvox was expensive, but had been shown to be as good or better than vancomycin in certain very serious infections. But Daniel lost this argument and vancomycin became the comparator antibiotic in this first clinical trial of 1212 in patients. The trial was completed in less than a year. Over 70% of the patients had MRSA as their infecting bacteria. Photographs taken of the skin lesions proved to everyone how seriously ill these patients were. 1212 performed at least as well as vancomycin in this trial and was associated with fewer adverse events than vancomycin. 1212 was ready for prime time.

After the data from the first trial of 1212 came in, David, Gerhardt, Wolfgang and the team convened to review everything. They then convened their scientific advisory board and presented the data. There was unanimous agreement that the data showed a favorable proof of principle for SW-1212 in serious skin infections.

The advisory board strongly suggested moving forward with the drug. But Swisbiotics could in no way afford the $70 million it would take to run the two phase III trials required to market 1212. They needed a partner and immediately set out to find one. They hired a bank, Citigroup, to help them. They presented at the annual JP Morgan conference in San Francisco. They met with investors and with pharmaceutical companies. There was a great deal of interest, but many balked at what they considered to be the limited commercial opportunity for the drug. By this time, at least one other IV and oral drug active against the MRSA superbug was going into late stage trials. A large number of IV drugs were already either available generically or would be marketed soon. Daniel pushed Swisbiotics to carry out its own market analysis to bolster its case that the drug would be valuable even with all the competition out there. But again, for lack of budget, they declined.

There was one company, Green Laboratories in New York, which was about to market an IV drug for MRSA. It needed something that could be used as a pill to follow their IV therapy and get patients home faster. SW-1212 would fit the bill. But they too were under financial pressure since their biggest selling drug, an anti-depressant, was about to lose its patent and become generic. They stood to lose $2 billion in revenue.

Green Labs finally agreed to an option deal. They would pay Swisbiotics $25 million up front. If SW-1212 could show that they actually had a pill form of the drug that would work well, that their new IV formulation was well tolerated and that they had a strong commercial case, Green would buy the company out at the end of one year. Green could walk away from the deal at any time during the option period, but had to make a decision yes or no by the end of one year. It was the best that Swisbiotics could do and the board was anxious for a deal. Daniel had pushed the board to provide an additional round of funding to Swisbiotics to increase its leverage in negotiating with Green. Daniel knew the management at Green very well from previous encounters with them. He knew how shrewd they could be. If they felt that they had you by the balls, they would never let go. The board refused. Green dragged out the negotiations until Swisbiotics was close to turning out the lights – and they accepted Green's offer *in extremis*.

Swisbiotics carried out trials of its new 1212 pill and showed that it was no more variable in getting blood levels than the IV drug. They showed that the IV formulation was well tolerated in volunteers and they carried out a more sophisticated market analysis suggesting that the drug would have peak year sales of $500-600 million globally in spite of all the competition. They accomplished all this within the first six months of the option period. Green Labs were pleased with the data, but refused to exercise their option. They dragged things out over the course of the year. Gerhardt and the board were getting more and more anxious. They could not understand why Green would not exercise since Swisbiotics had accomplished everything Green had asked for.

The answer was simple. At that time they had two years or so of market experience with their IV MRSA antibiotic and it was not good. Their sales were a miserable $10-20 million. Of course, Daniel always felt that Green knew a lot more about anti-depressants than it did about antibiotics and their flubbed handling of their own antibiotic showed that he was right (at least in his opinion). Daniel thought that Green was realizing that it was not an antibiotics powerhouse and that maybe it had made a mistake in taking the Swisbiotics product. Maybe Green felt lucky to have the possibility to decline the option. Gerhardt insisted that Green was, based on all his conversations with them, very positive and that they would almost certainly exercise their option at the end of the option period. Daniel was much less confident.

At this point, Daniel inserted himself in the thought process of the Swisbiotics supervisory board. He felt strongly that Green would ultimately decline their option. He wrote a memo to the board where he outlined his concerns and suggested that they develop backup plans to assure that they still had a way forward if Green backed out. Daniel suggested that they seriously consider going to the public markets in the US for funding for their late stage trials. He also supplied other options – but an initial public offering was his primary plan. His memo led to a long conversation with Deborah Cranston-Smith.

"Swisbiotics is a Swiss company. Why would we go to the US public markets?"

"Because the European markets are dead for IPOs – the US is recovering quickly. Also, European analysts don't really understand biotech and certainly not antibiotics. The US analysts get it and are very used to evaluating antibiotic companies."

"But we would have to establish a US affiliate."

"So?"

Deborah and the board refused to budge.

The end of the year came and Green declined their option saying that they were going to have to cut back rather than expand given the devastating loss of sales of their anti-depressant and the fact that their antibiotic was doing so poorly in the marketplace. Swisbiotics was almost back where it started the year before – but not quite. They now had a very strong package. They had completed all the additional volunteer trials they needed to show that they had a viable drug. They had a strong, new commercial analysis showing that the drug was valuable and could be a winner. Most importantly, they had all the drug powder they would need to make the IV formulation and the pills that would be required for their phase III trials and to get market approval. They also had negotiated a very strong deal with the FDA and Europe where they would be able to run a single trial in skin infection and a single trial in pneumonia and get market approval for both if both trials were successful. That means they could market in two indications instead of only one as had been planned originally. This also considerably strengthened their commercial position since they could now market in two kinds of infections as compared to just one. The only thing missing was the late stage trials and the actual formulations of intravenous and pill forms of 1212 that still had to be made. It would take them about six months from receiving funding until Swisbiotics could initiate trials.

Two years after Green declined their option, Swisbiotics was living on a shoestring of limited funding from their investors and a couple of Swiss government grants. They finally hired a new CEO. He negotiated a deal with a large, European venture capital group for 60 million euros of additional funding – more than enough to cover the two final trials needed before securing market approval. But Deborah was in a bind. Her fund was running out of money and she had been unable to fund another round. As the lead investor in Swisbiotics, her investment would be diluted and she would take a haircut if the new investor came in. She refused and the rest of the board meekly acquiesced. Eventually, Swisbiotics established a US affiliate and made plans for a US IPO. Whether they will be successful or not is still not clear.

Chapter 5

Marion

Daniel had been acquainted with Ludwig Hartz since Daniel's days back at the University in Cleveland. Ludwig was younger than Daniel by a couple of years, but had grayer hair, was taller and had a well-trimmed, pointed, white goatee. He was of medium build and about 6'2" tall. Ludwig had worked on enzymes of bacterial pathogens while he was a professor in Germany. He quickly went to work for Roche in Basel and was a victim of Roche's abandonment of antibiotic research in 1999. From Roche, he went to work at a small biotech centered in Munich but that maintained a research lab in Basel. Ludwig lived with his wife and son in Freiburg, Germany, not far from Basel. In this way, he didn't have to move. The biotech hired Ludwig and about 30 others from Roche in the hopes of identifying new antibiotics they could then partner with a large pharmaceutical firm.

Roman sent Daniel to Basel in 2004 when he was still at Virnuc to evaluate the company. With its cash from the public markets and the Gorman investment, Virnuc was searching for other companies and products to acquire. Daniel sat with Ludwig and his team and learned that they had only been working at the biotech for about four years. They had one possibly interesting antibiotic, but it was still only in the test tube phase. They had not yet done any animal model testing and little was known about potential for toxicity. Daniel also learned that Virnuc was interested because the biotech had decided that either the antibiotics group would partner with someone or they would close the Basel facility altogether since they were running out of money themselves. Their top investor had apparently informed Roman of this plan. The antibiotic was just too early for anyone to invest in it. It was the classic biotech story – 90% of them fail.

After the meeting, Ludwig called Daniel. He wanted advice on what to do. He was trying to come up with investors to spin themselves out of the biotech with some initial funding. The problem was that he wanted to take his entire group of 30 researchers with him and no one was willing to put up that kind of money for the one very early drug they had. They had a good conversation, but Daniel could offer no good advice other than to suggest that Ludwig lower his sites a little and think about starting much smaller. Ludwig wouldn't hear of it. He had worked with all these people for the last ten years. They were like a second family to him. He wasn't going to abandon them. Daniel and Ludwig agreed to stay in touch – and they did.

Daniel kept Ludwig informed about his own plans to leave Virnuc and to become a consultant. Ludwig announced that a small pharmaceutical company, Marion, located not far from their biotech labs in Basel, had hired his entire group. The company was publicly traded on the Swiss market and marketed a successful drug for a rare pulmonary disease. Marion wanted to move into the antibiotics field because they saw that resistance was creating a medical need. Marion saw that antibiotics was being viewed as a niche market and they thought that would be a perfect fit for them as a small, innovative and entrepreneurial company. They hired Ludwig and his entire group. All were relocated to nice new lab space in Basel. Ludwig invited Daniel to consult and advise on how to move their most advanced drug forward and to evaluate the other research projects they were pursuing.

Daniel arrived in September 2005 on the same trip as his first visit to Ullanger, also in Basel. He was given a tour of the facilities and met with both the CEO and the Chief Scientific Officer. Most of the scientists and administrators at Marion were ex-Roche employees just like Ludwig and his team. They had left Roche (were laid off) but managed to negotiate rights to a compound that Roche had decided not to develop. This compound, Hypergil, was the basis for forming Marion. It was carried into clinical trials by Marion, was successful at treating a rare lung disorder, and was sold at a very high price. When Daniel arrived for the first time at Marion, Hypergil was already selling $400 million per year and was projected to go to over $1 billion at its peak. Since marketing Hypergil, Marion was able to mount a successful IPO in Switzerland and was a publicly traded company.

The problem with Marion, Daniel noted immediately during his conversation with their management, was that they were a very small company with a large pharma mentality. They had all the metrics and bureaucracy Daniel associated with a much larger organization. They had a very slow decision making process with several layers of committees before approval for even small investments could be obtained. The other problem was that at the management level, they knew nothing about infectious disease or antibiotics. They seemed to be attracted to the area by some sort of romantic notion that they could be a white knight when all the other companies including Roche had abandoned the area. But their main source of knowledge and experience was Ludwig and his team. And Daniel was not sure that they were listening to Ludwig or whether they understood what Ludwig was telling them.

Ludwig and his team showed Daniel their most advanced antibiotic – the one Daniel had seen the year before. It was still just in the test tube phase of testing. The drug was interesting in that it was two different types of antibiotic, both known and marketed, chemically bound together to form a single molecule. They called it a chimeric antibiotic. They showed that each half improved the activity of the other half. They showed that each half bound to its own, different target in the bacterial cell. They also showed, most importantly for Daniel, that the chimeric antibiotic was active against bacteria resistant to the drug on one half of the chimera and the one on the other half. Against bacteria with both resistance mechanisms, the drug was less active than against susceptible strains, but was still more active that Daniel had expected it to be. This was probably due to the better target binding of the chimeric molecule.

Daniel had two concerns. First, the antibiotic was only active against Gram-positive bacteria and it would only be available by the intravenous route. With its structure, Daniel knew it would never be absorbed by the intestine and therefore could never be given as a pill. To Daniel this was a commercial disadvantage since there were already so many antibiotics available to treat Gram-positive infections including those caused by the MRSA superbug.

Daniel's other worry was toxicity. Since the chimera was made up of two different antibiotic classes, each with its own type of toxicity – mild though it might be – would the chimera be more or less toxic than either of the parent antibiotics? He suggested that Marion carry out a program to first determine what the therapeutic dose might be but more importantly to quickly establish whether the chimera would be safe at least in mice. He noted that it would be important for the new Marion group under Ludwig to have other molecules coming along behind this one since the chimera seemed very risky.

Ludwig and his team dove quickly into completing the experiments that Daniel had suggested. Within a month or two they were back on the phone with him with results. They had gone directly to testing the drug for toxicity in mice. One of the advantages they had in Marion is that the company had its own animal safety testing group and its own animal facility within its laboratories. So animal testing could be done quickly. They found that, as Daniel had feared, the chimeric molecule was toxic to the liver even at fairly low doses in mice. While this did not mean that the drug would be toxic in humans, it certainly raised the risk. Ludwig's team wanted to test the drug in another species – like the dog. Mice weigh about forty grams. Dogs weigh ten kilograms. There is a huge difference in the amount of drug required for testing in a few mice compared to that required for testing in dogs. This would require spending time and money.

They also wanted to continue to try and establish the therapeutic dose in mice in order be able to compare with the toxic dose they had defined. Daniel suggested that rather than using the formal model to precisely define the human therapeutic dose, they just use a lethal model of bacterial sepsis in the mouse to quickly but only roughly define a therapeutic dose. Daniel liked this model as it was quick, used a smaller number of animals and could give you a rough idea of how active an antibiotic would be in animals and later in humans. This would give them what would probably be an underestimate of the therapeutic dose needed for serious infections in humans, but if this dose was close to the toxic dose in mice, the drug would be very difficult to develop further.

Ludwig pointed out that in Switzerland, they were not allowed to study lethal infections in animals. The animals had to be sacrificed before they died. This made it very difficult to study the kind of infection Daniel had in mind – but he left it to Ludwig's group to figure out what to do.

Ludwig's team did as Daniel suggested but they had to watch the animals over a 24-hour period to be able to sacrifice those that were severely affected. They sent Daniel the data showing that the therapeutic dose in this model was close to the toxic dose in mice. Daniel thought that the drug could not be developed for human use unless the liver toxicity was somehow species specific. He advised dropping the program altogether. But Ludwig and his team were reluctant to kill the project. In fact, this was a familiar story for Daniel. Back at Penfrel he had to terminate a fair number of projects, but there was always pressure from the scientists to let the project live on another day. There were even occasionally surreptitious efforts at keeping projects going in secret hoping that some surprising positive result would get the project back on track. Those positive but surprising results never occurred in Daniel's experience.

So Ludwig's chemists worked hard to produce enough drug to test for toxicity in dogs. But then they found that when they gave dogs even relatively small doses of the antibiotic, the dogs would vomit. This was another problem Daniel had seen before. Dogs puke for no apparent reason. This was usually not a toxic effect of the drug, at least not for other animals or for humans, but seemed to be dog specific. But for Ludwig's chimeric drug, the only way they would be able to use a larger animal to test for the liver toxicity they had seen in mice was to go into monkeys. But they did not have monkeys in house – so this had to be contracted out. Again, this would mean more time and more money.

Daniel helped them find a contractor to do the study – in Japan. After six months, the Marion team finally got the result that all had feared. The chimeric antibiotic caused liver damage in both mice and monkeys. The drug could not be given systemically. But still, Ludwig's team held on.

The team said – what about *Clostridium difficile* (C. diff or CDAD to those in the know)? C diff was a common cause of diarrhea that occurred in patients during or after they had received another antibiotic. The organism, Clostridium difficile, would grow in the intestine when an antibiotic had suppressed the normal flora. C diff was frequently antibiotic resistant. It then produced toxins that poisoned the intestinal cells and caused them to secrete fluid and to die. The diarrhea could be bloody and could be severe enough to lead to surgery for removal of the colon or even to death. And, because C diff could form spores that could lie dormant for long periods of time, the disease tended to recur frequently even after apparently successful therapy. C diff was a big problem for the elderly and with an expanding elderly population being treated more and more aggressively with antibiotics in hospitals, it was a surging epidemic worldwide. Ludwig knew that the chimera could kill C diff but that it did not kill most of the rest of the normal flora of the intestine. He reasoned that since the chimera could not be absorbed, it would stay in the gut where it could kill the C diff and cure the disease. Daniel had to agree that this was a possibility.

Among many problems facing the team in their quest to make the chimera an anti C diff drug was that the animal models for C diff were not very good. They could predict whether a drug would work, but you could not use the models to accurately predict dose or dosing regime. This would have to be more back of a paper napkin sort of guess. The human gut has a volume of X. The concentration of drug needed is Y. The drug will be adsorbed to protein, to surfaces and to particles in the gut so we need to increase the amount by Z. But Ludwig's team persevered. Daniel helped by referring them to academic colleagues who could run animal model studies and carry out other experiments to help confirm that the drug would work. But, largely, this project went forward with little input from Daniel.

Daniel's main concern was that there were already several drugs marketed that could treat C diff. The problem was that some very severely ill patients did not respond well to treatment and that the relapse rate, especially among older patients, was high – around 25% or one in four. Some relapsing patients had to be treated multiple times before achieving a cure. Daniel counseled that their goal should be not to be superior or even non-inferior to current therapy in the first treatment, but to prevent relapse. This would be a higher hurdle and carry more risk since you could not know this definitively until you carried out large-scale clinical trials.

As Daniel reviewed Marion's overall portfolio of compounds in various stages of development throughout the company, he realized that, in spite of its risk, the one most likely to succeed was Ludwig's chimera for C diff. He also saw that in about five years, Hypergil would start to lose its market exclusivity and that generics would likely enter the market. This would decimate Marion's only source of revenue unless the company did something quickly. Daniel and Ludwig discussed this frequently during their biannual meetings. It was possible that the chimera could help fill the gap, but it would be unlikely to be able to make up for over $1 billion dollars in Hypergil losses at the time of the chimeric's launch. Daniel and Ludwig and their team developed a list of antibiotics in development by other companies, mostly in biotech, that Marion could consider licensing or purchasing outright. Marion had cash. They could afford such a deal. And they had a patent cliff staring them in the face. Since Daniel's other clients were developing several of the compounds on their list, this created a clear conflict of interest for him. He was forced either not to participate in any discussion of those compounds or to obtain permission from both parties allowing him to discuss them. Marion was happy for Daniel to be able to participate in these discussions and so were most of Daniel's clients. Daniel felt awkward about this but pushed ahead. He always felt that Henry Maloon, with whom he kept in close contact, would have disapproved.

Marion examined several of these opportunities in great detail. In every case Marion's lumbering bureacracy was either too slow to actually be competitive for any deal or it declined interest only to be outmaneuvered by circumstance. In one case, the biotech down the road from Marion in Basel, Swisbiotics, entered into an option agreement with Green Labs before Marion could get their act together. In another case, a Boston-based biotech carried out an IPO before Marion could decide whether they wanted to do a deal or not. After the IPO, their price became too high for Marion to pull the trigger. Daniel thought that this kind of perseveration was more typical of a large pharmaceutical company than a small and facile biotech and he attributed this to the Roche origins of most of the scientists and management at Marion.

Luckily, Ludwig's team had several other projects ongoing. In fact, Marion reminded Daniel of Penfrel when Daniel first arrived there. Marion probably had 10 different projects going and only 30 scientists – averaging 3 scientists per project. Daniel pointed this out to Ludwig who suddenly looked like a deer in the headlights. He agreed that they would have to prioritize and focus their efforts to be competitive.

One project involved a compound with an anti-bacterial spectrum very similar to the drug being developed by Swisbiotics, where Marion was several years behind. The drug could be used both IV and orally to treat pneumonia and skin infections. But Marion's drug was of a completely novel structure and its target for killing the bacteria was also new. This meant that there would be no cross resistance between Marion's new drug and other drugs either on the market or in development. Since no one knew whether any of the drugs in development would actually make it all the way to market, Daniel encouraged the team to pursue this new discovery. But, like everything else, toxicology was going to be important.

Ludwig and his team began the long process of using animal models of infection to define the therapeutic dose and dosing regime. At the same time, they began preliminary studies of its potential for toxicity. In this way, they hoped that, with a push, they could either kill the drug quickly or, if everything looked good, push ahead quickly and try and catch up with competition. The chemists were also working hard on back-up compounds with different structures from their lead compound in case the news from these early studies should be bad. The news was bad – but the chemists had a back-up ready to re-enter testing. They lost almost a year in this effort but were determined to push forward. Within six months the first answers on the backup were in – and they looked good. Ludwig then pushed the drug through the various committees at Marion and got permission to file an IND and start trials in human volunteers. They were behind Swisbiotics, but ahead of almost everyone else with a drug against the MRSA superbug that could be given in pill form.

In spite of all this good news, Marion still seemed recalcitrant to fully embrace antibiotics as a key area for the company. The two drugs being pursued by Marion that looked the most promising were, in fact, two antibiotics from Ludwig's team. Both homebrewed. Both targeted important areas of medical need and therefore important markets. Marion was still unable to bring any products in from outside the company to address their coming patent cliff. They only gave various antibiotic opportunities passing interest even though, if their two antibiotic development programs were to be successful, they would have a strong antibiotic franchise. A later stage product that could lead the franchise, at least in Daniel's opinion, would have been a logical step for Marion.

Daniel suggested, as he had from the beginning, that Ludwig and his team try and pursue a much more deadly bacterial target – the Gram negative superbugs that were emerging with ever-increasing frequency and that were resistant to almost all antibiotics on the market. This, of course, was a formidable task. Unlike the Gram-positive bacteria, the Gram negatives had not one but two cell membranes to keep toxins (like antibiotics) out. They also had a complicated and sophisticated set of molecular pumps to rid the cell of any antibiotics that might get through the dual set of cell membranes. Ludwig's team had been pursuing compounds similar to their Gram-positive antibiotic, but that worked on Gram-negatives. The problem had been that they could not get compounds active enough. They had proven that the problem was the pumps of Gram-negatives since bacteria where the pumps had been knocked out by mutation were very susceptible to the new drugs. They just had to find a compound that the pumps didn't recognize as well. With encouragement from both Ludwig and Daniel, the chemists never quit. They continued to try new avenues to modifications of their starting structure that might work better. And all this had to go on at the same time as the chemists were trying to make enough of their earlier compounds for early animal model testing, toxicology and other experiments.

About five years after having started their effort, in 2012, they hit pay dirt. They found a molecule where the activity was 10 to 100 times better than their early molecules and where, at least in the test tube, there was not evidence of human cell toxicity. At the same time, Marion's C diff chimeric antibiotic had advanced from human volunteers where it was well tolerated into later stage trials in patients with C diff colitis. With the discovery of a drug that could potentially treat Gram-negative superbugs and with the success of Marion's chimera C diff drug, the management at Marion finally seemed to recognize that their antibiotics team could be the future of the company.

In spite of this recognition of the potential importance of antibiotics to its future, because of the rapidly approaching loss of exclusivity on their major and billion dollar seller, Marion began to cut back on research and development. The antibiotic team was hit less hard than the rest of the company, but they still lost key members. Ludwig and his top chemist, both of whom worked together for the prior 20 years in Roche and later in Marion, were asked to take early retirement. A younger, and less experienced crew moved up to their management positions. At the same time, Marion finally seemed to get serious about bringing in an antibiotic product from outside the company that was closer to market than the two Marion drugs already in development. As of this writing, Marion still has not been able to pull the trigger on any sort of a deal to take them back from their cliff.

Daniel was saddened by the departure of his two friends and colleagues from Marion. He had grown close to them over the previous eight years of struggles together. But the three of them took satisfaction in that a number of compounds they worked on together were progressing in their trials and could someday become useful antibiotics for patients and physicians. Daniel retired from his consulting business before any of these compounds made it all the way to the marketplace, but he followed their progress assiduously. And he stayed in touch with Ludwig and other team members.

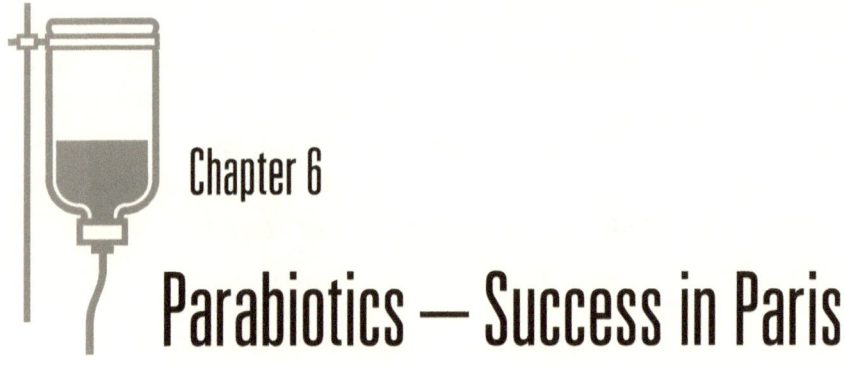

Chapter 6

Parabiotics — Success in Paris

In April of 2005, before Daniel headed off to Paris from Virnuc, he was invited to speak at a meeting on antibiotic development in Philadelphia. At that meeting, he met an old acquaintance from his Smith-Kline Beecham days at Brockham Park, Bran Wellman. Daniel had not actually spoken to Bran other than to say hello at various meetings for over ten years. His memory was that Bran was a ruthless manager in the style of Yosemite Sam from Gorman and Penfrel. Bran was tall, in his late 50s, balding and had a hooked nose. Daniel thought, on reflection, that he looked more French than like a Manchester-born Brit. Bran had moved on after the SKB fiasco to BAL Pharma, a German pharmaceutical company. BAL had just acquired a large French pharmaceutical company and was at that point one of the three largest pharmaceutical companies in the world. Its headquarters was now in Paris. Bran had been working at the Paris research site of the German firm before the acquisition – and he stayed in place. As Daniel and Bran were getting reacquainted and exchanging war stories, Daniel explained that he would soon be leaving Virnuc, but that he didn't know exactly what he would do afterwards. They talked about all sorts of possibilities in an informal way. But then Bran dropped a bombshell. BAL was going to close their Paris research site and was going to spin off their entire antibiotic discovery group. It would come complete with a number of compounds, all in various preclinical stages, but some, according to Bran, with real promise. Bran wondered, he said, if Daniel would be interested in participating somehow in the spinoff, either as a consultant or as a board member. This took Daniel aback, since he really still did not know what he would do after Virnuc. They agreed to keep in touch.

A few months later, when Daniel was in Paris, Bran asked him to stop by the Paris research site where the spinoff, now called Parabiotics, was located. BAL had given the site to the city to be used as a biotech research campus. The research site now had a few companies occupying BAL's old offices and labs. Bran took Daniel on a tour of the Parabiotics facilities and introduced him to the scientists and the one physician working there. The spin off had included about 35 BAL employees who were hired at their BAL salaries. Daniel was surprised to hear this since such high salaries were not usual for a biotech – especially not one just starting out. The site itself was showing its age and Daniel thought it could use a good cleaning and remodeling job. The carpets were getting threadbare. The labs were clean, but had accumulated indelible stains on the tiled walls and on the countertops and floors. But the scientific staff seemed bright, happy and excited to be part of this new venture and, for Daniel, that made all the difference.

While there, Daniel met their new CEO, Alistair Campbell. Alistair had wavy, reddish hair. Daniel thought he would look good in a kilt. He was about Daniel's height, was in good physical shape, had thin tortoise shell glasses, dark brown eyes and looked more like an intellectual than Daniels stereotypical image of a Scotsman. Daniel and Alistair had an old friend in common. Alistair had previously worked at Biogen. Daniel had worked closely with a wonderful scientist at the Institut Pasteur during his sabbatical who later worked with Alistair at Biogen.

Daniel, Bran and Alistair had lunch together at the cafeteria on the campus. It looked like the labs – in need of refurbishing. Daniel stuck with a baguette sandwich rather than the hot meals and salads that were available. Daniel had sent Bran his CV that was now over 20 pages long. Alistair obviously had read the CV since he asked specific questions about Daniel's previous work at Penfrel and Virnuc. He specifically asked about Daniel's work with the FDA both during his time at Penfrel and since. Alistair admitted that Parabiotics seemed ill placed to develop drugs for the US since they were lacking FDA experience. Everything that their team had done previously for BAL involved the European regulatory authorities and not the FDA. Daniel pointed out that the FDA, at least for getting trials started, was usually faster than Europe and could save them a month or two. This discussion led to the question as to whether Daniel would be interested in working with Parabiotics and if so how. Daniel said he would have to let them know since he himself did not yet know what he would do for his next venture. But between Parabiotics, Marion and Ullanger, the current pushing him into consulting was already strong.

Within a month or so, Daniel had made his decision. He called Alsitair who then arranged for him to meet with various board members. Daniel arrived at the BAL campus again in September just after his trip to Basel for meetings at Ullanger and Marion. Daniel discussed a possible role for himself with Bran and Alistair. Daniel's idea, based on his earlier discussions with them, was a rather varied role. First, he would establish a strong scientific advisory board for Parabiotics. He would then be an independent director on their board and would translate the opinions and advice of the advisory board for the board. At the same time, he would consult for Parabiotics on an ongoing basis on their various antibiotic discovery and development programs especially insofar as dealing with the FDA was concerned. He would also provide scientific advice when appropriate. Daniel admitted that this was a large brief, but he felt that this would be the best fit between himself and Parabiotics. Alistair was enthusiastic, but Daniel had the impression that Bran was less so.

As Daniel met with various board members, all venture capital or non-profit investors in Parabiotics, he was taken aback by their first questions beyond the usual "who are you" sort of introductory discussion. They all, independently, asked the same question in different ways. What did he think of Bran and his clinical colleague at Parabiotics, Syed Tengku?

Syed was of Malaysian origin and had obtained his medical degree at Cambridge, England. He had worked both in Malaysia as a general physician specializing in the treatment of TB and at BAL designing and supervising clinical trials of the BAL antibiotics. He was able to carry out his double life because BAL provided for five weeks of vacation to employees every year and Syed spent all five weeks working for essentially no salary in Malaysia. When they first met, Syed had spoken to Daniel about the large number of patients he was seeing in Malaysia with highly resistant TB infections. Daniel was very impressed with Syed's dedication to caring for these patients and found Syed to be an easy conversationalist.

When the board members first asked about Bran and Syed, Daniel was surprised. Only one of the various board members pursued his query in more detail. He wanted to know what Daniel thought of Bran's experience in manufacturing. Daniel answered that when he worked with Bran over ten years ago, he was not at all involved in manufacturing, but that perhaps that had changed during Bran's years at BAL. The same board member pressed Daniel about Syed's ability to deal with the FDA. Daniel already knew that this was an area of weakness for Syed and for Parabiotics, but he explained that he could help fill that gap. Daniel could not help but feel that there was more here than anyone was saying.

At the end of a long day, Daniel had a wrap-up meeting with Alistair. Daniel's first question was, "What is going on with Bran and Syed? Why did I get so many questions from the board about them?"

Alistair sighed. They decided to prolong their meeting and have drinks and dinner. This was going to take time. They headed out of the BAL campus – now called Paris Biotech – and took the metro, Paris' great subway system, to the center of town. They walked over to a quiet restaurant-bar in the Marais, near the Place Des Vosges, sat at a table outside and ordered scotch to start. Over their drinks, Alistair started.

"How well do you know Bran? Are you close friends?"

"I knew him during my years consulting with SmithKline Beecham, but have only run into him occasionally at meetings since then. We're more old acquaintances than close friends. You better just tell me what's going on." Daniel took a large swallow of scotch.

"OK. When Parabiotics was spun out of BAL, the most advanced product we had was PB-100. As you now know, PB-100 is a B-lactamase inhibitor paired, at least for now, with a penicillin like drug called tozime. The synthesis of PB-100 is difficult — its about 20 steps right now. But in order to carry on with our toxicological studies and eventually with human trials, we have to manufacture the drug in kilogram quantities. Bran said he would be able to supervise this. But so far, there has been nothing but problems and delays. The Board is getting tired of it and, frankly, so am I. And Bran's personality doesn't help. He is obviously disdainful both of the Board and me and he clearly does not believe we are capable of understanding the complexity of this manufacturing problem. The Board wants to let him go."

Daniel's eyebrows went up. Bran was Daniel's entré into Parabiotics. "I know something about manufacturing since I ran that effort at Virnuc. How about if I take a look at what's going on and get back to you with an opinion?"

"That would be great — but I think that this train has already left the station as you say."

"And what about Syed? Is it just his lack of experience at the FDA?"

"First of all, he is tainted by his connection with Bran. Bran argued strongly that he should be part of the spin out even before we were ready to go into clinical trials. Bran seems to have the opposite of the Midas touch as far as the Board is concerned. But there are other problems with Syed. He does not play well in the sandbox with others. He has to do everything himself. So he gets involved in the toxicology studies, in Bran's manufacturing projects and in plans for trials. But once he gets his teeth into something, no one else can participate. Its crazy. The other problem with Syed is that when it comes to speaking to potential partners and to new investors, both of which we will have to do in the next year or so, no one thinks that Syed is up to it."

Daniel was beginning to wonder what he was signing up for. "Is the Board ready to get rid of Syed too?"

"Not yet. I don't think so. But they clearly have their eyes on him and so do I, honestly. Your opinion here would be valued."

After their second drink, their dinners started to arrive. The rest of the evening, the conversation turned to Parabiotics' antibiotics advancing towards trials, and to home, family, travel and other more pleasant subjects. Daniel got to ask questions about various Board members. He also floated names for potential advisory board members. Daniel felt that Alistair was someone with whom he could work regardless of all the difficult issues they had just discussed.

Two months later, after Daniel and Sally had moved to Connecticut and settled into their new home, Daniel headed back to Europe. Parabiotics in Paris was his main stop, but he would also meet with his clients in Basel. He spent three days meeting with Bran, Syed and the scientists at Parabiotics. He also met with Alistair on several occasions. Daniel had drawn up a list of potential advisory board members and achieved consensus around a board of six with a couple of backups in case someone should decline. They agreed to have their first advisory board meeting early in 2006.

The Parabiotics scientists presented several projects to Daniel concentrating on their three most advanced programs, PB-100, PB-120 and PB-140. PB-100 was their B-lactamase inhibitor. This compound inhibited the enzyme that resistant bacteria produced to destroy penicillin-like antibiotics. Daniel had a good deal of experience with this type of compound since he worked on several while at Penfrel. The keys to success were that the compound had to be soluble so it could be administered to sick patients in an intravenous solution. It had to be safe since there were several other inhibitors on the market. And it had to hit more bacterial penicillin-destroying enzymes than the current marketed inhibitors. The reason for this latter requirement is that when the inhibitor is combined with the right penicillin-like antibiotic, the penicillin drug will be protected against more bacterial enzymes and would therefore be active against more resistant bacteria than anything else out there.

This sort of combination would target those difficult to treat resistant Gram-negative superbugs for which it is so hard to find good antibiotics. Such a combination drug, a broad-spectrum inhibitor with the right penicillin like drug, could therefore be a major benefit to patients and physicians. In the tests that Parabiotics had run so far, PB-100 appeared to meet all these criteria. It was very soluble. It seemed safe in the preliminary animal testing that had been completed.

Daniel raised a number of questions about PB-100 during these first meetings. One was about the penicillin like drug Parabiotics had chosen to partner with PB-100, tozime. Was it the right partner? The problem with tozime, Daniel knew, was that it tended to select for resistance. Even though the PB-100 part of the combination would prevent some of that – would it be enough? Wouldn't it be better to partner it with something like Penfrel's peracillin that was less likely to select for resistance and that would also bring activity against some Gram-positive pathogens that tozime did not have?

The scientists at Parabiotics had already looked at a number of potential antibiotic partners for PB-100 including Penfrel's drug. None were as active against the Gram-negative superbugs as the combination with tozime. Daniel said that he would like to spend more time looking at their data and that perhaps this was a question they could pose to their advisory board.

The other major question Daniel raised on PB-100 was about its manufacture. PB-100 was stuck where it was in development – waiting for its key toxicological testing – while waiting for enough drug to be made. Bran furrowed his rather bushy brows and got up to go to the whiteboard to explain to Daniel the current chemical synthesis and the manufacturing problems they were encountering. Daniel knew that there was a big difference between the way medicinal chemists think about making a compound and the way manufacturing chemists think about the same compound.

For medicinal chemists, almost anything that works is fair game as long as it won't blow up the lab. For manufacturing chemists, the process has to be able to be carried out at scale in large volumes and in a safe manner. Daniel also knew that he was not a chemist – but neither was Bran. They were both microbiologists with limited knowledge of chemistry. As Bran drew the 20 steps on several whiteboards, the room of scientists was silent. No one thought that this process would work to produce the kilogram quantities of PB-100 that were needed, except maybe Bran and Syed.

After this meeting, Daniel headed right for Alistair's office.

"Bran's plan for manufacture of PB-100 will never work. I'm working with a great manufacturing chemist who is now consulting. Lets bring him in – fast. Every day we delay is another day without enough PB-100 to go forward."

Daniel gave Alistair Clive Burn's contact information.

PB-120 was a drug that could only be made in pill form with a spectrum very similar to Swisbiotic's main product, but with a completely different chemical structure. Like Swisbiotic's compound, any clinical trials with PB-120 would have to be carried out in patients who were well enough to take pills. This would be a challenge, but Daniel thought it was quite doable. There were several other issues with PB-120 as well. First, PB-120 was very much like an antibiotic already marketed in parts of Europe and Africa. In fact, it was a natural product (made by a bacterium). It was a mixture of two different antibiotic molecules, PB-120-A and -B. These two molecules were extracted from bacterial cultures, purified, and then they were chemically modified to make the two ingredients for PB-120. Here, manufacture was not a problem since BAL already manufactured the marketed the related product, it was easy for them to supply the raw materials for Parabiotics. Parabiotics just had to have the molecules modified chemically in the right way – and this effort had been very successful so far. This was probably because they had had a good deal of help from the natural products chemists still at BAL.

The older antibiotic was known to cause a fair amount of gastrointestinal upset and even nausea and vomiting. There was also rare liver toxicity. Daniel knew from his experience with teracil at Penfrel that the animal models for the GI upset were not good. They would not be able to judge PB-120 from that point of view until they administered the drug to human volunteers. The liver toxicity, if it existed for PB-120, should be evident from animal studies.

The other issue for PB-120 was how much of which component should be given. The two components were metabolized at different rates, but the Parabiotics scientists had already determined that a certain ratio of the two components was best for activity against bacteria. Therefore, the dose in people had to take into account the differing rates of elimination of the two components to provide for the longest exposure to the correct ratio as possible. This was not going to be easy – but again, Daniel thought it would be possible. The scientists at Parabiotics were aware of this problem, but felt they could only make a guess at the ratio they would need before they tested the drug in human volunteers. Daniel agreed.

Daniel inquired about animal studies to determine the therapeutic dose for both PB-100 plus tozime and for PB-120. Neither had yet progressed to that phase. Daniel felt that animal models for PB-100 were not going to be needed. Since tozime had already been tested and marketed, the dose effective against susceptible strains of bacteria was already known for people. The PB-100 was just there to protect tozime from being broken down by bacterial enzymes. The question for PB-100 then was – how much do you need to protect tozime in the test tube and how can you dose it in humans to get this level at the end of the dosing interval – that is, right before the next dose. This, Daniel argued, could all be studied in the test tube. Once that was known, early studies in human volunteers would confirm how much PB-100 you would have to give to have that critical level of drug at the end of the dosing interval.

PB-140 was a program very similar to Marion's Gram-positive antibiotic program with a similar target in the bacterial cells. Parabiotic's drug could only be given intravenously, while Marion's drug was only given orally. PB-140 was also similar to Ullanger-Swisbiotic's drug in terms of the spectrum of bacteria that it killed. In fact, all of these drugs hit the same bacteria as Parabiotics PB-120. Luckily for Daniel, and consistent with his conflict of interest policy, none of these compounds shared anything like a similar chemical structure. Since all of these programs were fairly early in their development, no one knew whether any would succeed or not. Everyone was pursuing everything as if they only their own compound would succeed. Daniel agreed with this approach given the low success rates for drugs to get to market.

PB-140 had already gone through some of its animal toxicology testing and there was a potential for it to cause heart conduction problems. This was another similarity to the Ullanger-Swisbiotic compound even though the compounds were structurally vastly different. It looked to Daniel like PB-140 would pass or fail based on what would happen to electrical conduction in the heart in early testing in human volunteers.

All the other programs presented by the Parabiotics scientists were much earlier in their process compared to PB-100, 120 and 140.

After the entire series of meetings, Daniel met again with Alistair. Daniel said that their most important and most promising compound was PB-100 and that nothing should be spared in bringing this compound into clinical testing and then getting it to patients and their physicians. In Daniel's view, PB-100 looked like it would be very safe and, when combined with tozime or another penicillin-like antibiotic, would be active against a number of important and highly resistant Gram-negative superbugs that were plaguing patients and physicians. Daniel knew that the number of those infections would only grow while PB-100 was slowly wending its way to the market. Alistair commented that he had brought up the idea of a manufacturing consultant for PB-100 to Bran who refused. "You don't ask – you inform." Daniel replied.

Daniel was beginning to realize that Alistair was a remarkable person. Alistair was a businessman. He was not a scientist and had no formal scientific or medical training. Yet his grasp of scientific issues was impressive. He could also understand the medical needs around antibiotic resistance and put these in context of market opportunities. Alistair's presentations at advisory board meetings, at Board meetings, internally to Parabiotics' management, and everywhere else were clear, concise, organized and to the point. They also demonstrated his clear understanding of the concepts and issues in the world of antibiotics and how the Parabiotics projects could offer solutions. And, he had learned to speak fairly fluent French during his year or so commuting between Scotland and Paris. Daniel also realized that Alistair was going to be facing some tough decisions in the next 12 months.

Daniel attended his first Parabiotics board meeting in January. At the meeting, Bran provided a progress report on PB-100 manufacturing that was more like a lack of progress report. Clearly the board was not happy and neither was Alistair. Alistair raised Daniel's recommendation to bring on Clive Burns as a consultant to move the manufacturing along and the board unanimously agreed. Bran was overruled and he was not pleased. His brows furrowed – but he said nothing at the meeting other than to agree to contact Clive Burns. When Daniel approached Bran after the board meeting to discuss Clive and what role he could play, Bran brushed him off saying that he knew what to do.

At the next board meeting in March, Bran reported that a consulting contract had just been put in place with Clive and that there was yet another manufacturing delay for PB-100 related to a failure of one of the early steps in the synthesis. By May, Bran had been fired and he threatened to file suit. In France, firing someone was never easy and the company frequently ended up being sued in a special court called the Prud'homme. Employers faced a high risk of being overruled in the Prud'homme and they did what they could to avoid it. Everyone therefore tried to devise an agreement with the employee with this in mind. In Bran's case, no such agreement was going to be possible. Bran felt that his treatment was unfair and unjustified and he was going to pursue Parabiotics until Hell froze in order to extract some measure of proof that he was right and Alistair and the board were wrong.

Since Bran's main responsibility was directing the scientific groups, when Daniel arrived for his usual visit and the board meeting in May, he noted that the scientists were also angry and resentful at the loss of Bran and the way he had been treated. The scientific team meetings that

Daniel attended were wracked with discord. The biologists complained that the chemists were not producing enough compounds to drive the progress of the various projects. The chemists complained that for the compounds they were making, the biologists were keeping the data secret even though the chemists needed the data in order to design the next set of compounds to make.

Daniel spoke to Alistair. Alistair seemed unaware of these problems. But the project to find back-ups for PB-100 and PB-120 were at a standstill. The work on clearly identifying the mechanism of action of PB-120 was on hold for lack of compound. Daniel was at a loss for how to respond, and he wondered what his responsibility was for trying to make things right.

Daniel thought he knew what was happening. These conflicts probably always existed going back to the days of BAL pharma. Bran kept everyone in line and everything functioning with his iron fist. He simply would not tolerate this sort of infantile behavior. In his absence – it was the Balkans. Alistair had a good understanding of the personalities involved, but did not have the technical and scientific knowledge to intervene effectively. They needed to replace Bran somehow.

Daniel and Alistair worked on getting a list of potential candidates together, but it soon became clear that attracting someone competent and experienced to come and work in Paris was going to be a huge challenge. There were no good French candidates for the job. Most of the candidates Daniel could identify were already working at companies in the US and would be difficult to move. He called a few and ran into a brick wall.

There were no board meetings in France that summer, but Daniel came again in July to get updates from the various scientific teams before the entire company headed out on vacation. The peak vacation period in France is mid July to Mid August. When Daniel arrived, several key scientists were already gone on their 5 weeks of vacation. Some of them were going to take up to seven weeks. The only real progress to be reported was Syed's work on getting PB-120 and PB-140 ready for their first tests in human volunteers. Manufacture of pills of PB-120 and powder to be made into intravenous solution of PB-140 was well under way. The key toxicology data for both compounds would be complete sometime in October.

Even though Parabiotics had a drug formulations expert to deal with the manufacture of drug product – pills and IV solutions – Syed seemed to be in charge. He also seemed to be in charge of the toxicology even though Parabiotics had a good in-house toxicologist. Daniel worried about this since it meant that so much of Parabiotic's future was resting in the hands of a man who could not delegate. The good news was that Syed thought trials of both could start by the end of the year or early in 2007. Daniel raised the question as to which regulatory authority they would use – Europe or the US FDA. Europe, of course, Syed replied. But that will delay you at least an extra month Daniel noted. "Yes, but we know them well and feel more secure going through the European procedure."

In his usual wrap-up with Alistair, Daniel discussed the advantages and disadvantages of going through Europe vs. the US. They both agreed that they still needed to change the culture in Parabiotics such that they could deal with the FDA. They also discussed the fact that Syed seemed to be running drug product manufacture and toxicology even though he had little expertise in those areas. On the other hand, Daniel had been in the same position himself at Virnuc. The problem with Syed was that, unlike Daniel, he was unable to delegate responsibility to others. He had to be involved in every detail of these activities in addition to his clinical trial responsibilities and his job of working with the regulatory authorities to get the trials on track. That job alone was about to become much more than full time as they moved two drugs into their first human trials at the same time. But Daniel did not think that Syed could be moved.

Their discussion turned to replacing Bran. Daniel and Alistair had both called a number of contacts in Europe and in the US. Neither had candidates that were interested in the job or willing to move. They agreed that Alistair would have to engage a headhunter to fill this position, but that probably little would actually get done until the Fall. To Daniel's complete surprise, he found himself offering to take on Bran's job on a half-time basis until Parabiotics could replace him. In uttering those words, he had the feeling that someone else was directing his speech. Daniel admitted to himself that he had thought of this over the past several months, but he didn't remember coming to any conclusion – certainly not this conclusion. He thought that by working there half-time he could get some of the projects back on track and maybe even get the chemists and biologists speaking to each other. But he realized that with a half-time commitment, he would probably only get halfway there. That turned out to be 100% correct.

Daniel and Alistair agreed, pending agreement by the board, that Daniel would provide 10 days of work per month of which at least 5 days would have to be on site. He would have the title of Chief Scientific Officer and remain an independent Director. Parabiotics agreed to pay for an apartment in Paris as long as Daniel could show that it would be less expensive than the hotel they usually provided. Daniel asked for a fairly high reimbursement since he knew that a number of his clients would either terminate their contracts with him or reduce their use of his time given his commitment to Parabiotics - their competition. During the summer, Daniel and Alistair hammered out the necessary contracts and completed the official French government paperwork that would allow Daniel to work there part time.

Daniel was back in September to start his new position at Parabiotics. He informed his other clients of this change, and predictably, several terminated their contracts with him and others said little, but used his services much less while he worked at Parabiotics.

Sally came with him to Paris to look for furnished apartments. They found an one in the 15th Arrondisement. The apartment belonged to Simon, a friend of one of their close friends in Paris from their sabbatical days. He had been renting it out on a weekly basis to tourists. They liked the apartment – it was 70 square meters (630 square feet) and was on the 11th floor of an apartment building looking over the roofs of the city towards the Eiffel tower. The apartment reminded them of the one they had in the 11th arrondisement when they moved to Paris while they decided what Daniel would do with the rest of his life about two years earlier. This apartment was in a more modern building. It had a small but fully functional kitchen complete with espresso maker. The design, like their previous apartment, included a *cuisine americaine* or open design where the kitchen and living room were one large room. There were two large floor to ceiling *porte-fenetres* each with a small balcony looking out over the city. Sally and Daniel had seen a number of other furnished apartments, but the prices were either completely unrealistic or they were small, dingy, dark caves on the ground floor of some older building. Daniel offered to take Simon's apartment for a month at a time. Simon was thrilled with the guaranteed income and readily agreed to a discounted price that was considerably below five nights a month at Daniel's usual hotel. He did not insist on the usual (but illegal) one year security deposit, but did require a reasonable one-month deposit. He wanted to be paid in cash. Daniel and Sally asked few questions, but did confirm Parabiotics' willingness to reimburse them for the apartment. They moved in directly from their hotel and spent the next two weeks settling in to both work and apartment living in Paris for the fourth time.

Work was not easy. Daniel had four senior scientists reporting to him. Christian Mornay was the head of chemistry. He directed a small group of chemists working on the synthesis of Parabiotics compounds in large scale as well as a larger group of medicinal chemists who worked on Parabiotics very early programs as well as on finding backups for PB-100, 120 and 140. Christian was tall, thin and prematurely gray. He wore thick black-framed glasses that constantly slid down his relatively small nose. Christian had worked for about 10 years in Ohio at a large consumer goods company that wanted to break into pharmaceuticals. They had a small antibiotics research group where Christian was the head chemist. He spoke English fluently although Daniel preferred to speak with him in French.

Devin Redson was the head of biology at Parabiotics. He was shorter than Daniel, had thick, bushy blond hair and, like Bran, was from the northern part of England. He and Bran had been very close friends and Devin was the most resentful of the manner in which Bran departed Parabiotics. Devin, again like Bran, had a short fuse. He would release his anger in poison emails that he sent regularly with a single click to the entire management of Parabiotics. Yet, Parabtiotics was a small company. Devin ran a group of eight other scientists and his office was next door to Daniel's office and just down the hall from the CEO, the Chief Financial Officer and from Christian's office. Daniel could not understand why Devin insisted on responding in writing in emails when all he had to do was walk next door or down the hall to vent. Of course, to vent verbally, he might have had to soften his diatribe a little.

Devin was, without a doubt, a very clever scientist and was respected by everyone including Daniel. But he was secretive and closed. When Daniel asked about his unwillingness to share results with others on the team, Devin would say that the data or his hypothesis about the data had to be confirmed with additional experiments. But this could not work in a team that was trying to advance a project quickly. The chemists needed rapid feedback in order to decide which compounds to make next. If there was a promising compound in the last set, they might focus on that particular design. If there were toxic or inactive compounds, they might obtain key knowledge on what kinds of structures to avoid.

Daniel thought that this characteristic in Devin was the cause of most of the friction between Devin and Christian. Daniel spoke to Devin about his temper, his emails and his lack of cooperation with the chemists. His talks seemed to have little effect. Christian, on the other hand, was refusing to speak to Devin since he had basically given up. He believed he would never get the kind of rapid turnaround his chemists needed. Daniel brought the two of them together in his office. He explained that they would have to solve their problems together. Daniel made one of their key performance goals for the year the necessity for them to meet weekly to solve this problem of delays in getting data to the chemists. If they failed in achieving this goal, Daniel would decrease their bonus or their raise or both for the subsequent year. Daniel also took Devin aside and told him that each poison email would lead to the same kind of result. He encouraged Devin to vent to Daniel by phone or in person rather than to the entire company management. All of this resulted in an improvement in the functioning of various Parabiotics programs and the scientists working for Christian and Devin developed and strengthened their working relationships. But change was more difficult for Christian and Devin themselves – especially for Devin.

Malcolm Portman was the toxicologist for Parabiotics. When Daniel agreed to take over as a part time Chief Scientific Officer, he insisted that toxicology come under his purview. This meant moving Malcolm from Syed to Daniel. Syed was not happy about this, but he didn't object. Malcolm was thrilled because under Daniel, he would have much more responsibility. Daniel was happy to delegate the major responsibility for contracting toxicology studies and writing reports to Malcolm whereas previously Syed was constantly taking these over so he could have greater control.

Bernie Lowman was the head of microbiology for Parabiotics. He was in charge of the test tube tests for activity of Parabiotics compounds against key bacterial pathogens and also ran their animal model program in house. Bernie ran the contract work for both types of testing by contract research companies and academic groups as well. Bernie was quiet, easygoing, calm and never got into the Devin-Christian battlefield. His group provided data rapidly and openly to both the chemists and biologists without hesitation.

Within Devin's group, Marthe Montreux worked as a biochemist. She was thin, petite, wore thick-lensed thin-framed glasses and was clearly a woman in charge. She had two teenage children and was a single parent. Daniel speculated that this was the source of her attitude. She was the scientist who, more than anyone else, had driven the PB-100 project. She quickly became the project leader and took over the responsibility for coordinating activities across the team. Daniel spent more time talking to Marthe about progress on PB-100 and its backup program than he did with her boss, Devin. Marthe was smart, efficient and was usually around when Daniel needed to get an update on various aspects of PB-100 progress.

Working in France, Daniel soon discovered the joys of the French vacation system and their 35-hour workweek. France had passed a law restricting the workweek to 35 hours under President Mitterand in 2000. The idea was to increase employment. When companies balked at continuing to pay the same salary for less work, the government subsidized them. But hiring in France was so distasteful to most businesses that the 35-hour workweek had little effect on the unemployment rate. There were numerous problems. First, the salaries in France were relatively low because the overhead for employees was 150-200%. In the US, the overheads were closer to 25-30%. At one point, Daniel and Christian wanted to hire 12 chemists principally to increase their effort on the PB-100 backup program. They had something like 350 responses to their advertisements and about 325 of them were French. Daniel had had the same experience at the Virnuc labs in Paris. No other Europeans would come to work in France because the salaries were lower than in their home countries.

Then there was the challenge of problem employees. Because after the first three months, firing employees was so difficult and so expensive, companies hesitated to hire at all. This plus the lack of appropriately trained people in France helped keep the unemployment rate high.

Finally, the work schedule itself was dissuasive. In France – national holidays are not just one-day events. If the holiday should fall on a Tuesday, for example, the French would *faire le pont*. That is – they would make a bridge and take Saturday through Tuesday off. Because of the number of holidays that fell during May, employers could virtually write that month off in most years. Then there is the 35-hour workweek. What this meant, in practical terms, was that aside from the numerous national holidays in France, most employees could now take 7 weeks of vacation per year. What most of the scientists did at Parabiotics was to take 5 weeks during the year – usually a month in the summer and a week in the winter. Then they would take every other Friday off work. The great weekend exodus from Paris when Daniel lived there in the 1980s and 1990s was always Friday night just like it was for New York and Boston in the US. But in Paris of 2006 it was divided between Thursday and Friday nights. When Daniel would show up to work on Fridays, he would have two other Americans for company. But half of the scientists would not be there. That meant that half of the people with whom Daniel wanted to meet were not available on Fridays. He was never able to adapt to this peculiarly French work schedule.

While working at Parabiotics, Daniel got into the habit of stopping at a bakery on the way to work to pick up a sandwich for lunch. That way he could work in his office over the lunch hour. The two other Americans working at Parabiotics held a McDonald's Wednesday every week where they invited a few of the other employees to "McDo" for lunch. Daniel never bothered to go. The rest of the company headed over to the cafeteria on campus for a hot lunch. During Daniel's year as half-time CSO, there were two outbreaks of Salmonella food poisoning from eating at the cafeteria. But this did not keep people from eating there.

On his arrival as CSO in September of 2006, Daniel's first task was to get ahold of Clive Burns and bring him in for meetings with Daniel and Christian to sort out various manufacturing problems. The most acute problem was with PB-100 that was continuing to suffer delays and batch failures where their contractor would start to make a large batch of compound and then lose the entire batch at a later step. Bran had contracted with a chemical manufacturing company in India, Ran Ltd., to produce PB-100. Christian had prepared a detailed presentation for Clive that left Daniel in the dust. He left the two of them to work things out, but asked Clive to stop by for a summary of his thoughts before heading out.

Clive ended up spending an additional two hours in Daniel's office going over what he saw as the key problems with Parabiotic's Indian contractor. Clive pointed out that a detailed review of the raw materials that Ran Ltd. was using showed that they were significantly inferior to the ones that Christian's group had used to carry out the reactions successfully at Parabiotics. So the first fix he suggested would be to insist that Ran start with raw materials that were at least as pure as the Parabiotics chemists use. He also pointed out that while he thought that the Ran chemists were actually very competent, in Clive's experience, when dealing with manufacturers in countries like India and China, it was best to have your own chemist on site for at least two weeks out of every three months. In this way, he said, you could nip things in the bud before they could cause a big problem like the loss of an entire batch. Finally, he said that Parabiotic's strategy of letting the Ran chemists develop a new and shorter pathway would probably not work. Their expertise was more in getting the reactions you give them to work – not in discovering new reaction pathways that might be shorter or more efficient. Clive wanted to put a couple of Christian's chemists on this problem right away. Daniel thanked him for his clear input.

Clive left saying that he would send a written report in a week. Daniel ran back to Christian's office to debrief. Christian was excited about his meeting with Clive and said that for the first time he might be seeing a small light at the end of a long tunnel. He already knew which chemists he was going to put on the problem and who he would be sending to India in the next few weeks. Daniel suggested that for the first time, they take Clive with them.

Daniel then headed down the hall to Alistair's office. He explained how excited both he and Christian were about working with Clive and how they felt that things might actually finally start working for PB-100. Alistair smiled, but didn't say much. Daniel understood. Seeing is believing.

In fact, Clive quickly took control of the PB-100 manufacturing process at Parabiotics. He gathered a team of four chemists and put them on research to improve the scale-up synthesis of the molecule. Of course, Christian was working closely with Clive to get this done since Clive was still only a consultant. This posed a problem since Parabiotics only had about 12 chemists total and now only eight were available to work on the other Parabiotics programs. Within a few months the process research team had made two significant breakthroughs. Clive and the process research team leader headed over to India to Ran's plant to get the process in place there. They stayed about two weeks until they were sure that the reactions the team had put together in the lab would work in India. They did.

On his return, Alistair offered Clive a full time job at Parabiotics. Daniel was surprised and pleased. It was Daniel who first recommended Clive and it was Daniel who forced him down the throat of Bran and eventually it was again Daniel who got Clive more deeply involved in the PB-100 manufacturing problem. When Clive accepted Alistair's offer, Daniel was even more surprised. Daniel knew that Clive had a successful consulting business that he would have to give up if he worked full time at Parabiotics. But he did. Daniel asked Alistair how he convinced Clive to make the move. "I made him an offer he couldn't refuse." Daniel didn't want to even think about what that might be – but solving the PB-100 manufacturing problem would make Clive more than worth his weight in gold.

PB-120 was running into problems. Or at least Syed thought so. As it was a drug with two components, PB-120-A and –B, animals absorbed much less PB-120-B than humans did. This meant that it was difficult to get high enough levels of B in animal tissue to be sure the drug would be safe in humans. Although the regulatory authorities hadn't ever stated that they recognized this as a problem for continued development of PB-120, Syed was very concerned about the issue. He discussed his concerns with Daniel. Daniel thought that they should simply ask the authorities their opinion. If they wanted more or different studies, they would say so. Syed was reluctant to discuss anything with the regulatory authorities. He pushed for an entirely new set of safety studies using PB-120-B in an intravenous formulation. But these studies would have to be carried out in monkeys since the IV form was not tolerated well by dogs. The bill for Syed's proposed studies was almost $1 million and the results would not be available for another year. Alistair and the board agreed to allow Syed to conduct the studies over Daniel's objections. The data from these studies were reassuring about the safety of B, but Daniel was never sure whether the studies were truly necessary or not. Did they reassure potential partners on the safety of PB-120? Did they provide data that the regulatory authorities would have required anyway? Daniel never knew.

At the end of 2006, Parabiotics began receiving serious inquiries from medium and large sized pharmaceutical companies about their antibiotic programs. As Daniel had predicted, most of their interest centered on PB-100 as it was the only drug that targeted Gram-negative superbugs that were viewed as the medical need of the near future. This increased the pressure on Alistair to find a way to replace Syed. Alistair asked Daniel to be present at meetings with potential partners to provide a more experienced clinical and scientific view of programs and to answer questions when Syed stumbled.

Alistair and Daniel discussed Syed in several meetings. Alistair had decided that Syed was just not going to be able to lead Parabiotics' discussions with potential partners and new investors. He felt they needed someone more dynamic, more experienced, and more in tune with the American way of doing things, including working with FDA. Daniel pointed out how he had suggested that they open discussions with the FDA several times to Syed, but that he had made no progress.

Things were more difficult after Bran's departure. Syed used to report to Bran. Rather than have Syed report to Daniel, Alistair had him report to Alistair directly. Now Alistair was planning to hire someone to replace Syed. Alistair's idea was to try and keep Syed at Parabiotics while hiring a more senior clinician. Syed would then report to this new Chief Medical Officer but would still maintain some of his responsibilities depending on what the new hire would want to do. Alistair added a CMO to the CSO recruitment the headhunter had already undertaken for Parabiotics. Daniel worried that Syed would see this as a demotion.

Daniel had staff meetings every month. His four direct reports and Syed were always invited. Starting in early 2007, there were two agenda items under recruitment – CSO and CMO. Progress on both fronts was discussed every month. Although the discussion of the Chief Scientific Officer position was frequently animated with many suggestions for candidates or comments about candidates selected to come in for interviews, there was silence on the Chief Medical Officer recruitment. Syed, in particular, said nothing.

Although there were a number of candidates that had been identified for the CSO position and a few would be interviewed in the first part of 2007, there was no progress on the CMO front. There were no candidates to be found in Europe and none of the attractive candidates in the US were willing to move to Paris. So many companies had abandoned antibiotic research and so many of their clinicians had gone on to other things, that there were really only a few viable candidates anyway. During one of Daniel's many meetings with Alistair on recruitment, Daniel suggested that Parabiotics establish a clinical development center in the US. This would serve two purposes. First, a US affiliate could communicate directly with the FDA and secondly, it would provide access to the best antibiotic developers in the world for their CMO recruitment. Alistair presented this idea to the Board in early 2007. They agreed with the plan. Alistair decided to wait until they hired someone and let that person choose the site for the Parabiotics clinical development facility in the US.

Alistair interviewed the first candidates for both positions in early 2007. As the year wore on, none had yet appeared at Parabiotics to be interviewed by anyone else until late Spring of that year. The most exciting candidate for Daniel was his old friend and colleague from Penfrel, Richard Noland. Richard had stayed on at Penfrel after they had decimated the infectious disease group. He took on a job as vice president of protein research at the old Hemotech facility in Boston. But his first love was still infectious diseases – so he agreed to look at the CSO job at Parabiotics in Paris. Alistair wined and dined him and took him to a soccer game at the Stade de France on the outskirts of

Paris. Richard spent a day at the facility in Paris talking to Christian, Devin, Clive, Syed and Daniel. He wrapped up the day with Alistair. Alistair was enthusiastic and optimistic. Daniel said that he doubted that Richard would move. He suspected that Richard was looking at other possibilities. After almost two months of fruitless negotiations, Richard informed Alistair that he had accepted a job at Ullanger's new Boston research facility where he would head their anti-infectives effort. Daniel always thought that Richard would only be happy at a large company. Small companies had too much risk and too few capabilities for Richard. But this turn of events put Parabiotics back to square one in their search for a CSO – they had no backup.

In the meantime, Bernie Lowman announced to Daniel that he would leave Parabiotics in one month. He had accepted a higher paying job at the French affiliate of a larger research contract organization with laboratories in the same campus as Parabiotics. Daniel tried to negotiate with him offering a larger salary and other enticements, but Bernie was committed. Daniel pressed him for his reasons, but never got a clear answer. He always suspected that the sour relationship between Devin and Christian and the constant infighting played a major role in his decision to leave. His departure would mean the Daniel would have to play a bigger role in the microbiology group adding to the already high pressure of his part time job at Parabiotics.

Daniel quickly got permission to hire a headhunter to replace Bernie. By the end of a month of searching, the headhunter had come up with three candidates. Two were French and Daniel thought that neither was particularly well qualified. The third was American and very experienced. But why would he want to move to France? The microbiology job at Parabiotics had to stay in France – it would be impossible to move since the rest of the group was all French. Daniel brought in two candidates for more extensive interviews, including the American, Bob Jones. He was about 45 years old, wore thick glasses, had dark hair with the beginnings of a balding head and sported a full, bushy beard. Bob had worked at a large pharmaceutical company in North Carolina for a number of years, then at a biotech in the Washington DC area. Pullman Pharmaceuticals in the UK had just acquired that biotech. Bob was out of a job. But at his two previous positions, he had taken one antibiotic through all of its microbiological testing all the way to the marketplace. He knew the contractors, the academic contacts and he knew his way around a microbiology laboratory. Bob was just what the doctor had ordered for Parabiotics – with one hitch.

During his interview, Daniel probed Bob's enthusiasm for moving to France. Daniel had made this move many years ago and knew that the adjustment would be hard. Bob's wife did not work and she didn't drive. They had two high school age girls both with learning disabilities. Daniel thought that moving his family to France was a disaster in the making. He tried to talk Bob out of the idea. But Bob insisted that his family were all very enthusiastic as was he. Daniel had no other viable candidate and Bob had the ideal experience for the Parabiotics job. Bob was offered and accepted the position. He moved with his family about two months later.

The Parabiotics' headhunter came up with a new CSO list of possible candidates by July. Among the names, Daniel recognized one in particular that he knew from Gorman, Trevor Malcolm. Trevor was a microbiologist at Gorman. He was very familiar with test tube and animal testing and had been involved in several antibiotics where drugs had made it all the way to market. He had experience, therefore, in the research required to get a drug into clinical trials, to support it through its development and to continue that support in the market place. Plus, Daniel liked him personally. He said he was willing to move back from Gorman's labs in the US to Europe. He still had aging parents in the UK. His wife was still working in the US but was excited about the possibility of spending part time in Europe. Trevor visited Parabiotics at the end of July – just before the worst of the French holiday season. Even then, only a few of the key people were around. Syed was there but Devin was not. As it turned out, Trevor and Devin were old buddies from Gorman somehow – so Trevor knew what he was getting into on that front. Alistair offered him the job.

After three months of negotiations on salary, stock, vacation, and everything else under the sun, Trevor accepted. He started in October. Daniel could go back to his consulting practice and leave the politics of being CSO to Trevor. Alistair, Daniel and Trevor agreed on a transition of several months where Daniel would come to Parabiotics for one week every month to work closely with Trevor around the various Parabiotics projects until Trevor felt on top of everything. In fact, within six weeks, Trevor was managing everything including the Devin-Christian feud. Daniel asked him how he did it. Trevor said that he had just taken Daniel's cue of using performance ratings to push Devin into more appropriate behavior. Plus, Trevor's relationship with Devin gong back many years probably helped – or at least that's what Daniel told himself. There were still rare poison emails from Devin's office, but they were rare and they resulted in an unpleasant and immediate response from Trevor.

Trevor asked Daniel about Bob Jones. Trevor was having difficulty keeping Bob focused on the tasks he had prioritized. Daniel acknowledged that this would be a constant problem, but that with supervision, Bob was quite manageable. They also discussed Bob's personal situation. Daniel knew little about this – but he imagined that it could not be good. As things ultimately turned out, Bob's wife was furious that she had been forced to move. She was completely isolated in their apartment outside of Paris. There was no convenient public transportation nearby. She did not speak French and she did not drive. Their two daughters were enrolled in the American school not far from the apartment and were picked up by bus every day. But they too were isolated, were not making friends and were not happy. Daniel and Trevor learned all of this during an awkward dinner that Trevor arranged as a get-to-know-you sort of outing among the three couples, Trevor, Bob and Daniel with wives. Bob's wife seemed to be seething throughout the evening.

Shortly after the dinner, Trevor arranged for Bob to move back to the US where he would work exclusively with outside contractors doing microbiology work for Parabiotics. Trevor himself would supervise the French microbiology group supporting early projects.

By the late Spring of 2007, the CMO interviews were picking up. There was only one candidate from Europe that made it through Alistair's interview. But Daniel quickly saw that although he had a fair amount of experience in clinical trials of antivirals, he knew nothing about antibiotic development. Daniel pointed out that he would face a steep learning curve if he came to Parabiotics and would therefore not have the presence and knowledge to go toe to toe with regulators, investors and potential partners. Alistair quickly dropped him. Two other candidates came from the US. One was a consultant who was very hesitant to go back to full time work with a boss and a board of directors and he rapidly took himself out of the running. That left one candidate who was working for a large pharmaceutical company in the Philadelphia area.

Barbara Mankowski had recently taken an important new anti-fungal drug through all its trials to the marketplace and had solid experience developing antibiotics as well. She was short, blonde, a little on the stocky side, had thick glasses in clear plastic frames and sported a warm smile that emphasized her maternal side. Daniel had not met her before. But he called some mutual friends that had worked with her in various capacities. They described her as focused, a strong leader, but also as someone who would "take no prisoners." She stood up for what she believed was right, would fight and argue and work to get things done the way she considered was correct, but would yield to reasonable argument according to her colleagues. Daniel spent several hours with Barb both during her interview in Paris and during a second interview in New York where they met at a well-known steak house. Daniel was convinced of her abilities and he thought that if anyone could deal with Syed, it would be her. Daniel would be proven wrong on that front.

Like Trevor, Barb started working for Parabiotics in October of 2007. Her first six months were crushing. She was almost never home. Her husband, a radiologist who worked at an academic hospital outside of Philadelphia, was used to Barb's compulsive work habits and tolerated her absence and the moodiness brought on by the extraordinary work pressure. Barb had to identify space for her yet to be established clinical development group. She was tasked with establishing the Parabiotics affiliate in the US. She had to get all three of the Parabiotics clinical projects in hand. To accomplish the latter, she had to rely on Syed. But there, she ran into a complete roadblock.

By the time Barb and Trevor arrived, discussions with Green Labs had progressed considerably. Green was discussing licensing PB-100. Under the proposed terms, Green would provide an upfront payment and would pay for half of PB-100's development costs. Daniel thought that the upfront proposed was reasonable. Alistair and the Board thought that if Parabiotics was going to have to shoulder half of the development costs, that the upfront payment should be more generous.

Negotiations dragged out through the Fall. Alistair invited Daniel and the Parabiotics management team to an early Christmas dinner at a well-known Parisian restaurant in early December. During dinner, he disclosed that he had sent a counter-proposal to Green Labs earlier that day. It involved an upfront payment of $150 million. Daniel was floored. PB-100 had not yet completed its earliest trials in human volunteers. Although manufacturing was going much better than it had a year earlier, it was still not straightforward. Daniel did not believe that Green would ever pay that much for such an early stage antibiotic. Between the cheese and desert courses Alistair was interrupted with a phone call. He raised his bushy eyebrows as he picked up his cell phone. Daniel knew it was Green. Alistair walked out of the restaurant into a cold, clear night to take the call. He came back looking somber and sober. Then he broke into a large, toothy smile. "They bought it! $150 million upfront!! Time for champagne!" No one had a bigger smile than Daniel. But Daniel knew that either Green had no idea what they were doing, or they were desperate for a new antibiotic and they wanted to secure the rights to PB-100 before anyone else – or maybe both things were true.

With the loss of chemists, the backup team that had already focused solely on finding a superior PB-100 was foundering. The chemistry was extremely difficult. And they had no expertise on their team in the area of structure-based design where you use the actual 3D structure of the target protein to help design or optimize drugs to bind or inhibit it. Daniel and Christian put together a plan that relied on a chemical strategy to come up with a better PB-100. They then worked out how many chemists it would take to drive that plan. In their plan they also included a chemist with expertise in structure-based design and a contract with a company that could provide the structures they would need. The problem was that this would all take a large investment by the Parabiotics investors. But Daniel thought that with the new license agreement from Green and the progress in manufacturing of PB-100, the mood on the board would be to support this effort. The Board clearly recognized that PB-100 was the most valuable asset they had and having an improved compound that had not been licensed by Green would be the way to further leverage this value. He was right. The Board approved the project.

Christian and Daniel set about to hire 12 more chemists. There were two problems. The first was the peculiarly French problem that Daniel had encountered at Virnuc. Only French people will take jobs working in France. Once again, for twelve open positions, there were over 300 applicants, 95% of whom were French. Christian and Daniel went through almost 30 candidate interviews to select the twelve to whom they would offer a position. All accepted.

That led to the next problem. They were out of lab space for chemists. Each team of two chemists required one chemical hood. All the lab bench space and, more importantly, chemical hoods, were already in use at Parabiotics. Christian and Daniel had to search out more space on the Paris Biotech campus where Parabiotics was located. They identified space in a nearby building. But it would take another six months before that space was usable. In the meantime, the chemists that started joining Parabiotics would have to work in shifts if they needed time in the chemical hoods. This was not exactly the French way – but everyone was so excited about the project that there were virtually no objections.

Barb, Alistair and Daniel met at Parabiotics facilities in Paris in November. Barb was just finishing up a punishing week trying to catch up on the Parabiotics clinical development projects. She did succeed in getting materials that were kept in the large file drawers in the hallway adjacent to Syed's office and to everything that was available on the Parabiotics computer servers. But Syed kept much of the material on his own computer or in his office and he refused to meet with Barb. Alistair was caught in a trap. Syed was in the middle of preparing the reports from the first trials of PB-120 and PB-140. Only Syed had access to the data coming in from the testing centers since he was the Parabiotics physician of record for the studies. Alistair was reluctant to simply fire Syed because he thought that Syed was competent and he felt that he risked losing a great deal of time while Barb started totally from scratch if Syed refused to share anything with her before departing.

Syed refused to speak with Barb for several reasons. First, it turns out that he was, quite simply, a misogynist. Female secretaries were OK. Even female clinicians were OK as long as they reported to Syed. But Syed reporting to a woman Chief Medical Officer? Never. And this was a topic he would not discuss either with Daniel or with Alistair. Alistair and Daniel tried, in turn, to cajole, convince and threaten Syed – all to no avail. So Alistair decided to divide the baby. They would give Syed one of the projects as his own, PB-120, and for this, he would report directly to Alistair. Barb would have everything else in clinical development. This seemed to function until the PB-120 project started to fall apart.

The first problem with PB-120 was the mixture of the two components. It turned out that their human volunteer studies showed that people and mice were completely different in their absorption of the two components from the intestine. While mice absorbed more of the B component, humans absorbed more of the A component. To achieve the correct balance in the bloodstream, they would have to reformulate tablets with a different ratio of the two drugs and restudy the new tablet in a few human volunteers.

Syed then elected to carry out a trial of the new PB-120 formulation in patients with pneumonia. Daniel had reviewed the protocol for the planned trial before Barb was ever hired. He asked for major revisions such that the trial would be carried out in patients with serious infections. Daniel was aware that the regulatory agencies both in Europe and the US were demanding that drugs for pneumonia be studied in more seriously ill patients than had been the case in the past. Also, a study in seriously ill patients would allow for an easier transition to studies in patients with serious skin infections.

Syed never responded to Daniel's detailed notes on the PB-120 trial design. It turned out that the trial designed by Syed did not even allow for the calculation of disease severity. Alistair approved the trial design without knowledge of the correspondence from Daniel to Syed. He was anxious to have a success. In retrospect, Daniel thought that Alistair had calculated that Syed's trial would be quick and provide positive results.

Daniel and Barb reviewed the PB-120 pneumonia trial results in mid-2008. Daniel showed her his email exchange with Syed requesting modifications. Syed had ignored Daniel's advice thinking that he could get the trial done more quickly if he avoided the issue of disease severity. Daniel and Barb both agreed that this was true, but suspected that most of the patients in the trial had such mild infections that they would be unable to conclude anything from the trial. This would mean that they would have to run a separate trial in patients with serious skin infections before going into late stage trials that could lead to market approval. In turn, this meant that PB-120 would be delayed by at least one year while another trial was conducted. They presented their conclusions to Alistair. He was furious – but Daniel wondered whether he was more angry with Syed, Daniel or with himself.

The result was that Syed retired amicably from Parabiotics. Daniel kept running into Syed during his work with other clients in Europe. Syed seemed to drift from one company to another staying for a year or two at each.

In contrast, Daniel found working with Barb was like a dream. Many of his struggles and worries over trial design were lifted. He could actually collaborate with someone who knew what they were doing. Barb was a good listener, but had her own ideas many of which were good ones.

At the same time, in mid-2008, the Parabiotics board hired a bank to look at selling Parabiotics to a large pharmaceutical company. There were several challenges to overcome. First, none of the drugs had completed trials in infected patients except the PB-120 pneumonia trial that was useless. Then there was the fact that Parabiotics' most valuable asset, PB-100, was already licensed to Green Labs and there was not yet a backup ready for the clinic. Finally, Alistair himself was problematic. He wanted to found a new, small pharmaceutical company and build it himself. He saw Parabiotics as the next Cubist.

The board persevered and brought in several banks to present at board meetings. Most of this was beyond Daniel's understanding, but clearly the investors had an idea of a minimal sale price that would bring an acceptable return on their investment. They ended up picking Goldman as their bank. But by the end of 2008, little progress had occurred. Alistair was dragging his heals. He was more interested in pursuing cheap compounds at other companies that he could bring in to Parabiotics than he was in selling the company. But you can only drag venture capitalists around for so long. The Board laid down the law to Alistair. Get behind the sale or move aside. He got behind the sale.

Things started to move in early 2009. By then, Barb finally had her arms around all of Parabiotics clinical programs. She became the leader and spokesperson that Alistair and the board had imagined. Barb and Clive were clearly key elements in the ability of Parabiotics to align itself to be acquired by a large pharmaceutical company. Clive had overseen a remarkable turnaround in the situation with PB-100 manufacture to the point that the cost of goods was predicted to be quite reasonable at commercial scale. Barb had led the clinical trials of PB-100 defining the dose that would be required for therapy in humans and showing that there would be no safety concerns for the drug. She also revamped PB-120 development. She essentially discounted the pneumonia study conducted by Syed and started a new study in skin infections. There was good progress on finding a new compound that would be superior to PB-100 although there was still not a candidate for the clinic. This progress and the efforts of Goldman led to keen interest by virtually all the large pharmaceutical companies with active antibiotics programs. Glaxo, Merck, Astra-Zeneca, Gorman, Pullman and Green were all reviewing Parabiotics data with an eye towards making some sort of offer.

When the problems of heart conduction with PB-140 became clear at the end of early trials, after a discussion with the FDA, Barb recommended that further development of the drug be halted. Daniel was awed by the ability of Barb and the Parabiotics management to take this step since it was so unusual in biotech to kill a project even when the data indicated that was the right choice. Each project added to the value of the company, at least in principle, so management and boards were loath to eliminate projects entirely. But Parabiotics went against the grain and did so based on their own analysis of their own data.

As the merger and acquisition process got started, the Parabiotics board, guided by Alistair, made two important moves. They committed to investing additional monies over the next year such that they would never be caught short of money during their negotiations with potential partners. This meant that they could negotiate from a position of financial strength. They also moved to make sure that in the event of a buyout, all the Parabiotics employees would get something from the stock options that they all owned. This meant that the investors had to agree to give up on some of their return. To Daniel's surprise, the investors agreed to a very reasonable package for the employees in spite of the cost to them. He never forgot the empathy and foresight of the Parabiotics investors and Alistair's leadership.

Among all the pharmaceutical companies expressing serious interest in Parabiotics, it was Pullman Pharmaceuticals that seemed the most eager from the start. Pullman was the largest pharmaceutical company in the world. The merger of two large, European pharmaceutical companies had formed Pullman several years earlier. They were one of the few large companies to still have an active antibiotics program and they were headquartered in Britain. The biggest problem for all the potential partners was what to do with Green Labs since they held the US rights to PB-100, the most important Parabiotics asset. Of all the companies meeting with Parabiotics, Pullman was the only one, at the end of the day, willing to split the pie with Green. This meant that there was only going to be one suitor and that they could, to a large extent, dictate deal terms to Parabiotics. Pullman would be forced to cede US sales to Green while Pullman would take sales in Europe and the rest of the world including Japan. Parabiotics' plan to have two competing suitors was dead. Pullman was in a very strong negotiating position. But Parabiotics was not KO'd. Thanks to their investors, they had enough money to keep going through all the negotiations and well into another year of work in case things fell apart with Pullman. Goldman made sure that Pullman knew about Parabiotics' financially secure position and of the attitude of its investors. They were not going to accept a lowball offer.

During all these discussions in 2009, PB-120 was being studied in patients with serious skin infections. PB-100 was just starting its trials in complicated intra-abdominal infections like appendicitis, gallbladder infections and abdominal abscess. At the same time, a trial in complicated urinary tract infection was beginning. Pullman would have to acquire Parabiotics quickly in order to control the data that would come out of these trials and to prepare for the trials that would be used to register these drugs for market. Those trials could start as early as 2011.

Based mainly on the success of PB-100 in its early trials, and with a nod to the possibilities for PB-120, Parabiotics was acquired by a Pullman for almost $500 million at the very end of 2009. This was remarkable since neither of the two assets that were acquired in the deal, PB-100 and PB-120, had completed their first well-designed trials in patients although such trials were in progress at the time of the acquisition. No data from these trials were yet available. The only data on PB-100 and PB-120 were from animal models, from safety data in animals and from human volunteer studies. This, of course, was the magic of antibiotics. From these early data, the ultimate clinical success of a drug can be predicted with great accuracy at least as far as whether the drug would cure infections or not. The risk of a rare but important side effect was ever present. In the case of Parabiotics, PB-100 went on to be an important antibiotic for physicians and patients. PB-120 never made it to the market because it was not as good as its comparator antibiotic in its skin infection trials and there were problems with nausea and vomiting. Neither problem alone would have kept it from proceeding, but the two together combined to kill the program.

Pullman Pharmaceuticals

The acquisition of Parabiotics by Pullman was bittersweet. Parabiotics was an undisputed biotech success story. The investors and management did very well financially and Pullman and Green acquired a nice antibiotics franchise. Not only that, but there was the promise of actually being able to do some good for patients and physicians in need. At the same time, most of the employees at Parabiotics lost their jobs immediately upon closure of the deal. Many would still either be unemployed or would be working in other areas four years later. A few, like Clive Burns, Barb Mankowski and Trevor Malcolm were kept on for six to twelve months to allow for a smooth transition from Parabiotics to Pullman. Barb led the data analysis for the trials of PB-100 in mid-2010. The PB-120 data would not be available until the end of that year and by then Barb would have taken a job in another large pharma company.

Daniel lost his Parabiotics consulting position immediately upon deal closure and he was singularly uninformed about the progress of PB-100 and PB-120 throughout most of 2010. Daniel tried talking to Barb and later to colleagues at Pullman, but no one was talking. He felt isolated. It was as if he had lost contact with close family members.

2010 was a tough year for Daniel in that it took him the entire year to build his consulting portfolio back to where it had been before Parabiotics. Many companies had either terminated their contracts with him or simply used his services less or not at all for the years when he was so deeply enmeshed in Parabiotics. When Daniel was fully free from Parabiotics, it took him time to reclaim old clients and find new ones.

Pullman also had a tough year. They had an antibiotic of their own in development. It was a fluroquinolone type drug, like ciprofloxacin (Cipro). Their drug was more active against the Gram-positive organisms than Cipro and other drugs like it, could be taken once a day and, at least in animals and in all the human trials until 2010, had looked very well tolerated. The Pullman clinicians and scientists all met over a period of several months to design the trials they would carry out to win final regulatory approval to market their drug, florafloxacin (Floracin), worldwide. They had high hopes for this antibiotic. The fluroquinolone market was big – several billions of dollars globally. These were some of the most commonly used antibiotics among physicians and patients. There was resistance, but Floracin was active against many of the bacteria resistant to other quinolones. Pullman knew, though, that there had been several spectacular failures of other antibiotics in the same class and that these failures were occurring only after large number of patients had been exposed. This means that if there were some problem of rare toxicity, they might not see it until these final trials or maybe not until the drug was marketed and hundreds of thousands or millions of patients were exposed.

One of the key questions during the Pullman meetings was – what dose? Should they use the highest dose they had tested in earlier trials because it would cure the most resistant infections and would be less likely to select for resistance to Floracin itself? Or should they hedge their bets and use a lower dose because of worries about possible rare toxicities they had not yet seen. The project leader at Pullman argued strongly for using the higher dose. "Floracin is clean. It shows nothing in animals. Not even the usual quinolone toxicities were seen in our test tube tests or in our animal studies. And we know that resistance to Floracin will come more quickly at the lower dose. There is no scientific rationale for choosing the lower dose," he said. He won the day. The higher dose was chosen. The trials had started in early 2009.

About halfway through the trials, they noticed that some patients were having trouble with drops in blood pressure for no apparent reason other than the antibiotic they were getting. Since half of the patients were receiving Floracin and the other half were getting Cipro, and since they didn't know who was who since the data were blinded, they worried. Luckily, they had thought to put in place a Drug Safety Monitoring Board to review safety data at the halfway point of the trial. The DSMB could look at the data first in a blind fashion, just like the Pullman scientists, but also they could break the blind while keeping the data secret from Pullman if they needed to examine things further. This is exactly what the DSMB did. They found that Floracin was associated with sudden and sometimes serious drops in blood pressure in about 5-10% of the patients where this was not seen in the patients getting Cipro. The DSMB recommended that the trials be stopped. Pullman did as they requested and Floracin, after having spent over $100 million in research and development, was trashed.

The failure of Floracin, one of the many late stage quinolone antibiotic failures, led to a good deal of backstage second-guessing. Would this have happened at the lower dose? The Floracin project leader eventually left Pullman to found his own company carrying out the kinds of studies that would provide a strong scientific basis for the choice of antibiotic doses in early and late stage clinical trials. He never forgot his Floracin experience. And neither did Pullman.

Luckily, Pullman did not let this experience dissuade them from antibiotics as evidenced by their purchase of Parabiotics. In 2011 Daniel ran into the Pullman scientists at a meeting in Europe where they were presenting publicly for the first time the results of the trials that Barb (with a little help from Daniel) had put together while at Parabiotics. The data were astounding. They showed that PB-100 was as good if not better than the best, last line antibiotic available against even the most resistant superbugs infecting patients in the study. This was true both for intra-abdominal and for urinary tract infections. Daniel saw Barb at the meeting and they both celebrated at drinks and dinner one evening. They talked about what was going on at Pullman and when the final trials required for registration would begin. But neither Daniel nor Barb had any idea and the Pullman folks weren't talking. Daniel asked what happened to the PB-100 backup program that seemed so promising when Parabiotics shut down. Again Barb had no idea.

The next day, Daniel ran into the Pullman scientists at lunch. They sat together and Daniel had an opportunity to ask a few questions. They slid into a discussion of what the final trials could look like. Daniel was suggesting a radical approach where PB-100 would be studied in the treatment of superbug infections where the pathogens were resistant even to the last line antibiotic on the market. The Pullman clinicians looked at him quizzically – as if he had just stepped off a spaceship or something. But they were intrigued enough to sign Daniel up as a consultant. Daniel would finally get to see what they were doing with his favorite antibiotic candidate from among all the ones he had ever worked on.

In his first meetings with Pullman in London they presented their plans. Daniel's first question was "What have you been doing for the last two years?" But they sidestepped the question by getting down to their design for their late stage clinical trials. They had met with the FDA and with European regulators on several occasions to review trial designs. The trials they envisioned were in urinary tract infection, intra-abdominal infection and in hospital-acquired pneumonia where the latter would start later and where the design of that trial would be the most controversial. But their basic plan going across all these trials and extending to a fourth, uncontrolled trial, was to show that PB-100 was effective against tozime-resistant Gram-negative infections.

Daniel was surprised and disappointed, although this is what they had hinted when he ran into them at meetings several months previously. "But several marketed antibiotics already work to treat tozime-resistant infections. PB-100 plus tozime will kill many Gram-negatives that are resistant even to our last line antibiotics the carbapenems. Why can't you focus on those infections?"

"Because we don't think we will be able to recruit enough of those kinds of patients into our trials as currently designed."

"Right. Agreed. Throw out your designs. Lets think of a small trial where you enroll patients where carbapenem resistance is highly suspected. They all get treated with PB-100-tozime. Maybe, if we're lucky, 10% will actually have carbapenem-resistant infections when the cultures come back from the lab. That 10% becomes your analysis population. You only have to show that the antibiotic works in those patients. The rest become part of the safety database – to show that the drug is safe. According to the new thinking of the regulatory agencies in Europe, you probably only need to study 30 patients with carbapenem-resistant infections to get approved. Your safety database could be as small as 300 patients. You get on the market quickly. You charge a high price. And you carry out additional trials while making money in order to expand the kinds of patients you can treat."

"Patients with what kinds of infections?"

"All comers. Urinary tract, intra-abdominal, pneumonia, skin, whatever. The agencies are talking about these kinds of trials. Let's show them one. What have you got to lose?"

"And what will we do about controls? The agencies always ask for controls."

"Good question. I'm not sure they will insist on controls given the difficulties you will have in recruiting these patients. I suggest we put an uncontrolled design together and then discuss controls if the agencies insist. We can see what they have in mind."

Daniel didn't see or hear from the Pullman scientists for another two years. He stayed in touch with Pullman's head of clinical development who was an old friend and colleague. But he would never get into the details of Pullman's plans. Daniel did keep an eye on the website, clinicaltrials.gov, where all the ongoing trials and their designs are posted. He found that Pullman had essentially ignored his advice and stuck with their original plans. One good thing he saw was that for their pneumonia trials that were started two years after their original trials, they had completely ignored the FDA's impractical and infeasible requirements and designed their trial based on Europe's more pragmatic designs. In other words – "FDA – here it is. This is going to be a great new antibiotic active against resistant pathogens infecting American patients. Take it or leave it." Although Daniel was disappointed that they had rejected his trial design advice, he was pleased they had taken this attitude with the FDA.

In fact, Pullman became one of the leaders in getting the regulatory agencies to agree to the kinds of trials Daniel had suggested. But, ironically, they elected not to pursue these designs for their own antibiotic. Daniel thought that the problem was one of timing. When PB-100 was ready for pivotal trials, the regulatory agencies were still in the thinking stage about the kinds of design Daniel had in mind. The conservative regulatory affairs folks at Pullman didn't want to push the agencies at the time, thinking that they might lose the argument.

In 2013, Pullman called Daniel back in as a consultant. This was four years after they had purchased Parabiotics. They wanted him to review their antibiotic research portfolio. The year before, Pullman hired a new CEO. In early 2013, under pressure from the loss of sales due to generic intrusion into their sales of key products, the CEO announced that antibiotics would not be a priority area for the company and that investment in the area would be scaled back. Daniel saw this and wrote a scathing editorial about the policy. In his editorial, Daniel emphasized the important role that Pullman had played in getting the FDA and European regulators to prioritize and streamline antibiotic development. He also discussed the recent meetings taking place where insurers both in Europe and the US were seriously discussing higher prices for antibiotics. Again, Pullman was playing a leading role in these discussions. How could they lead the way for other companies in the field of antibiotics and at the same time threaten to pull out entirely? This led to an exchange of letters with the CEO and numerous phone calls with various antibiotics research heads including his old friend and colleague. The end result was that they invited Daniel to their headquarters in London to review their progress on various antibiotics research projects. Daniel assumed that this was to demonstrate that in spite of the CEO's statement, they still had good things to come. And, in fact, it was true.

Daniel's major interest in going to Pullman was to see what had happened to the old Parabiotics backup program to PB-100. The first question he asked when he arrived was, "What have you been doing with our PB-100 backup program for the last four years? We had already made good progress when you purchased Parabiotics? I've seen and heard nothing since. Where are my B-lactamase inhibitors?" While they did show him what they had been doing, and while Daniel was pleasantly surprised at the additional progress they had made, he was not sure what had taken them so long. Daniel pressed them for an explanation. The Pullman scientists looked uncomfortable. Finally, the head of the scientific group said something like, "We have recently overcome some impediments to clear strategic thinking, here." Daniel finally understood. Pullman had prioritized other programs over the PB-100 backup. The people responsible for that decision were no longer at Pullman. Daniel supposed that this meant that Pullman had finally understood the error of their ways.

But Daniel was very concerned that Pullman would join the ranks of Roche, Lilly, Abbott, Penfrel, Wyeth, Bristol-Myers-Squibb, Johnson and Johnson and Pfizer in abandoning antibiotics research altogether. His meetings in London did not reassure him even though he was very impressed with their antibiotics research portfolio. He felt that their CEO should also be impressed. Six months later, Pullman announced that they would close one of their anti-infective research centers in Malaysia. Was this the beginning of the end for Pullman antibiotics?

Daniel placed a call to his colleagues at Pullman. "What is going on? Is there any significance to the fact that Pullman is closing its Malaysia research facility beyond just that?" His contact just referred him to news on the web summarizing recent statements by Pullman's CEO. Pullman was, like many other large pharmaceutical companies, faced with generic intrusion on the market for their largest selling drug – for the treatment of ulcer disease and heartburn. Their revenues were sinking faster than the Titanic and Pullman management was scrambling to cut costs. The CEO, during their quarterly update for the markets, said that they would slow their investment in infectious diseases research to focus on other areas that they thought would be more profitable like cancer. After reading this, Daniel called the head of the scientific group in London. "What about PB-100-tozime? Is that still on track? What about the PB-100 back-up program?" He was assured that everything was still on track and was told not to worry. But, if everything was still on track, how was the CEO saving money? Daniel didn't believe a word they said. But he believed that the scientists and clinicians at Pullman truly believed that somehow they could be de-emphasized with their programs and personnel intact. Daniel's experience at Penfrel taught him that wishful thinking was just that.

Within six months, Pullman announced that its antibiotics unit was up for sale and had been for months with no serious buyers. Pullman itself was being pursued as a takeover target by at least two other large pharmaceutical companies both of whom had given up antibiotics research long ago. Daniel's worst fears seemed about to come true. Would these promising antibiotics really be put on the shelf?

Then Daniel read that the US partner of Pullman, Green Labs, was going to ask the FDA to approve PB-100 based on the phase 2 trials Barb had carried out for Parabiotics – four years after the data had become available. The FDA indicated that it would consider this and set up an advisory committee meeting. Daniel watched the meeting cheering and swearing as if he were at the superbowl. The committee voted to approve PB-100. This antibiotic would be available to Americans before Europeans or anyone else on the planet. The label would be restricted to indicate that the antibiotic should be reserved for those patients with infections where the treatment choices were very limited. But this approval would set an important historical precedent for antibiotics and would show that the conservative approach of Pullman was antiquated.

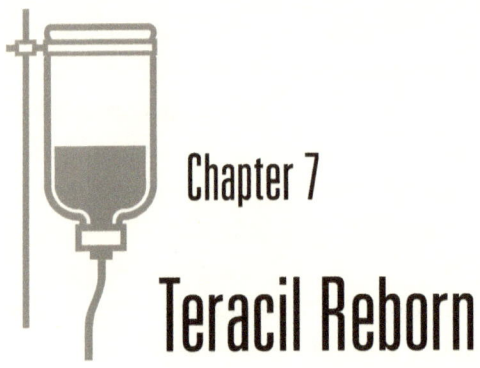

Chapter 7

Teracil Reborn

When Daniel was still working at Penfrel, while attending an infectious diseases meeting in San Francisco, he was shocked to see a paper on a molecule that looked just like teracil – almost. It differed by only one or two carbon atoms. Daniel approached the Penfrel chemists to ask how this could possibly happen. Presumably, all of the potential chemical modifications of teracil should have been covered by the Penfrel patent – therefore this new molecule should either not exist or it should only exist in Penfrel's own laboratories. The chemists just shrugged their shoulders. The Penfrel patents did not cover the molecule, then called CL-219. Daniel was furious – but there was nothing anyone could do. The new molecule had been synthesized at a company that Daniel had never heard of, but that had been founded by an old academic colleague of Daniel's in Cleveland. The company was called C-Biotech and was located in a new set of research buildings not far from the Cleveland University campus where Daniel used to work.

At the same meeting, C-Biotech announced a new collaboration with Gorman Ltd who had paid an upfront fee of around $25 million and had rights to all of C-Biotech's molecules including CL-219. CL-219, though, was clearly inferior to Penfrel's teracil in that it did not kill the Gram-negative superbugs as well as teracil. Daniel dismissed it as an irrelevant close call.

Several years later, just as he was finishing up at Virnuc, Roman asked Daniel to evaluate C-Biotech and CL-219 as a possible takeover target. Daniel found that since he had first seen CL-219 in 1998 or so, C-Biotech had gone through a series of partners. Gorman pulled out within a couple of years, but Daniel could find no rationale for Gorman's decision. Then C-Biotech partnered with Bayer who pulled out when they abandoned antibiotic R&D altogether. Following that, they partnered with Pullman Pharma, but that deal had just fallen apart. In spite of all this, CL-219 had managed to get through its early trials in human volunteers, but had not progressed beyond that point when Roman sent Daniel to Cleveland.

In looking at their data, Daniel discovered that CL-219 was much better tolerated in human volunteers than teracil. He knew that clinicians were not particularly concerned about the nausea and vomiting caused by teracil, especially when treating highly resistant infections where there were few other choices for therapy. But he also discovered that CL-219 might be absorbed orally. This might be a game changer. This would provide for a new antibiotic that could be taken in pill form that would hit at least the MRSA superbug and maybe a couple of others. There was only one pill form of antibiotic on the market at the time that could treat MRSA. It was hugely expensive and, in spite of that, was selling well. Why? Because there was sometimes little choice. But the C-Biotech data suggesting that they might have a pill form came only from animal studies. This would have to be confirmed once again in human volunteers. Daniel was optimistic, based on the animal data that this would work.

During the two days Daniel spent reviewing all of the data on CL-219, he had time to interact with a number of C-Biotech's scientists, but more importantly, with the CEO and their head of business development. Frank Thomas had been the CEO of C-Biotech since it was formed. He was a big man with a big smile, light curly hair and bright blue eyes. He seemed friendly and affable. Frank had led the company into all of its partnership deals and was still standing after they had all fallen apart. Daniel pressed him as to the reasons for the loss of Gorman and Pullman given that CL-219 looked like a reasonable potential product. Frank shrugged his shoulders as if to say – who knows? In their discussions, Daniel began to suspect that part of the reason for the failed deals might have been Frank himself. Frank seemed to hold himself in high opinion – not unlike Roman – and Daniel could imagine that, also like Roman, he would not take the advice of others, even those with deep pockets, very well.

Daniel provided a positive report on C-Biotech and CL-219 to Roman. But, at the time, Roman was unable to come up with the kind of money that Frank was demanding. And C-Biotech had enough money in the bank from their failed Pullman deal to carry them through at least another year or so.

Daniel was surprised to get a call from Frank a few months after his visit there on behalf of Virnuc. C-Biotech was getting ready for their first trial in patients. Daniel was starting to consult and Frank wanted him to work with C-Biotech to help them get ready for the planned trial. Daniel was more than happy to do this since it gave him a chance to get back to see old friends in Cleveland. Sally was thrilled by this as well.

The work started with a review of the C-Biotech data demonstrating the dose required to cure infections in standard animal models of infection. Daniel's old friend John Anders had carried out these studies. As Daniel would have expected, the studies were well done and the data were strong. They showed that a human dose of 200 mg per day would be sufficient and that the drug could be given as a single daily dose.

Then Daniel reviewed the toxicology data and found that even the irrelevant toxic effects seen in animals with the tetracyclines (teracil and CL-219 were both tetracycline-like drugs) occurred at doses two to four times higher than the dose effective in treatment of infections. This was better than he had seen with teracil, which had already made it to the marketplace.

Daniel then reviewed the data from the trials of CL-219 in human volunteers. At a single dose of 200 mg, there were essentially no side effects. Even during a trial of 200 mg per day for seven days there were still no problems. In fact, C-Biotech had raised the dose as high as 600 mg for seven days and only just began to see some nausea at that dose. There were no toxicities like decreased blood counts, liver damage or anything else that Daniel could see. CL-219 would be effective and would have fewer side effects than teracil. Finally, he reviewed the data in human volunteers for the pill form of CL-219. The first data had just become available and these trials were continuing at the time that Daniel conducted his review. The data so far showed that C-Biotech would have to administer about 600 mg of CL-219 orally to get the same blood levels of drug as they saw with 200 mg given intravenously. But the 600 mg dose had only been given once so far. They would have to go through a week of dosing to be sure that it would be well tolerated and then they would have to go to higher doses to make sure that there would be some sort of safety window.

After spending two days reviewing all these data, Daniel met with the C-Biotech Chief Medical Officer, Ron Alston. Ron was busily writing the protocol for a trial of CL-219 in patients with serious skin infections. He had designed the trial to allow up to 14 days of therapy with the antibiotic. He had also allowed patients to switch from intravenous therapy to pills whenever they were better enough to eat normally and they no longer had fever. The trial compared CL-219 to the only marketed oral therapy for MRSA superbug infection – the one that sold for the highest price of any marketed antibiotic and that would attain $1.5 billion in sales globally before it lost its patent protection. CL-219 would therefore be a potential replacement for a drug selling well over $1 billion. Of course, this first trial in patients would not be enough to get CL-219 approved by the regulatory authorities but it would set up the trials that would lead to such approval.

In their discussions, Daniel offered several thoughts to Ron. He expressed his dismay that the volunteer trials had only treated subjects for 7 days while the proposed trial would allow up to 14 days of therapy.

Even with the lack of side effects at the dose to be used, 200 mg once per day, this would require the regulatory authorities and C-Biotech to accept some risk. Ron replied that he had raised this issue with Frank and his management at the beginning of the volunteer trials, but that Frank was unwilling to pay the additional money involved in a longer trial in volunteers. He agreed with Daniel's concern, but felt that if the regulatory authorities balked, there was nothing he could do anyway. Ron did say that with CL-219's favorable side effect profile, he thought it likely that the authorities would allow the trial to proceed as written.

Daniel suggested that they allow physicians to test the bacteria they isolate from patients to be sure they would be susceptible to CL-219. But C-Biotech had not carried out any of the studies that would be required to allow microbiology labs to do this routine testing. They did have supplies and data that would allow a central lab to test all the bacterial isolates in batch at the end of the trial, but these data would not be available to the physicians treating the patients in the trial. Some testing was required because it would tentatively establish what level of antibiotic would be required to tell if an organism was susceptible or resistant based on the results of this first trial. These data would then be refined during the larger trials that would be required for registration. Although the design using a central lab to test bacteria only at the end of the trial was allowed by regulators, Daniel himself had never run a trial in that way. He had always provided test data to physicians treating patients to allow the doctors to make the most appropriate decisions regarding whether or not the patient should continue in the study or not. In the C-biotech design, half of the data would be unavailable to the physicians.

Daniel suggested that Ron consider treating patients with more than one dose of CL-219 in this first trial. What if the bacteria began to become resistant to the 200 mg dose after a few years on the market? If the drug was so well tolerated, would it be better at least to check out a higher dose to see if a higher dose was even an option? Ron agreed that this would be a good idea and said he would follow up with Frank. But a week after Daniel returned home he received a note from Ron saying that Frank was insisting on a single dose, 200 mg per day, in the trial. Frank was confident that the dose would work and would be well tolerated and he did not want to risk side effects at the higher dose nor did he want to spend the additional money another dose group would require.

As it happened, after this single visit to Cleveland, the rest of Daniel's work for C-Biotech was done from his home in Connecticut. There were no more trips back to see friends while working with his new client. Ron would send Daniel versions of the investigator's brochure, the clinical trial protocol and the protocol to be used by the central microbiology lab to test the bacteria obtained from the patients. The various versions would go back and forth until C-Biotech was satisfied that they were ready to submit everything to investigators and to the regulatory authorities so they could get approval to start the trial.

Daniel was stuck on one point that he considered important. In the documents, the regulators required that companies justify the dose they were proposing to use. Daniel had written this entire section for Ron at Ron's request. But Ron kept sending him back versions of documents with a different version – always the same. In this alternate version, the dose justification was based entirely on a comparison between CL-219 and teracil, the marketed drug from Penfrel. Daniel knew that the regulators would not accept this as a dose justification. They needed to see the data testing CL-219 directly in key animals models just as John Anders had done for C-Biotech. The section Daniel had written was based on those data and never mentioned teracil at all. After several exchanges of versions, Daniel called Ron.

"What is going on with this dose justification section? I keep rewriting it and you keep changing it back to something the FDA will never accept?"

"I know, I know. Its not me. Its Frank. He wants the comparison approach in the documents to convince thought leaders and investigators that CL-219 is better than teracil."

"What does Frank know about the FDA? He has no business putting his little advertisement in one of the most important sections of the documents required by the FDA for approval of the trial. He can put the comparison with teracil anywhere in the document he wants – but not in the dose justification section. The FDA won't accept it. They will just send it back asking you to revise it and this will delay you at least four to six weeks."

Daniel never heard back from Ron before the documents were submitted to the FDA. He wrote Ron asking for a copy of the documents sent to FDA. The dose justification section as written by Frank was what was submitted. One month later, C-Biotech received a fax from the FDA with their comments on the CL-219 submission. One of them was a request to provide a justification of the dose chosen for the trials based on the standard animal model of infection. The FDA would not allow the trial to proceed until the justification for the proposed dose was complete. With that, Ron was finally able to replace Frank's section with the one written by Daniel and, after another two months, the FDA allowed C-Biotech to proceed with the trial.

Daniel did not hear from Ron or anyone at C-Biotech until after the trial had been completed and the data had been analyzed, more than a year later. The results were clear. CL-219 performed at least as well as the comparator antibiotic in serious skin infections and was clearly superior to teracil in terms of side effects. Most importantly, the CL-219 pills were also well tolerated even at a 600 mg per day dose. Ron presented the data at the big infectious diseases meetings that year. The presentation was covered by the press and made quite a buzz at the meeting.

Just after the presentation, C-Biotech and Ullanger announced that they had entered into a collaboration to develop and to market CL-219. This would be C-Biotech's fourth deal with a large pharmaceutical company. Daniel hoped that it would be the charm. With the conclusion of this deal and the Ullanger team behind them, C-Biotech no longer needed Daniel's help. He had done nothing for the prior year anyway. A few months later, Ron left C-Biotech to join another biotech. He had had enough of trying to deal with Frank.

During the following year, Daniel heard little about CL-219. He just assumed that the two companies were getting everything ready for the final set of trials required for approval of the antibiotic for the marketplace. Daniel knew that C-Biotech had not invested in the manufacture of enough drug to support these trials and he supposed that the delay was a combination of time for manufacture of drug powder, of the intravenous formulation, of pills and also to allow for multiple meetings with the FDA and the European regulatory authorities in order to negotiate the design for these final trials.

But after over a year of waiting for the companies to announce the beginning of their trials, Daniel received a note from one of his colleagues at another company suggesting that Ullanger had pulled out of the collaboration. Daniel had seen nothing suggesting this might be true. He had not heard from anyone either at Ullanger or at C-Biotech. To try and confirm the rumor, Daniel searched through the annual report from Ullanger and their regulatory filings. He found one sentence in a quarterly update to the SEC noting that they would dissolve their relationship with C-Biotech. That was it. No press release. No other public announcement from either company. At this point, another friend of Daniel's who had worked at Gorman, had taken the job of Chief Medical Officer for C-Biotech. Daniel gave Cal Gallagher a call.

"Is it true? What is going on there?"

"Its true. How did you find out?"

"I heard from a friend and then I found this single sentence in an SEC submission by Ullanger a couple of weeks ago."

"OK. Just between you and me, this can only be a good thing. First, there was little for me to do here with Ullanger running the show. Now maybe we can get a decent design together."

"You mean you don't have a trial protocol yet? Its been over a year."

"You're telling me! Ullanger couldn't decide on anything to save their lives! I submitted at least three different synopses for a straightforward trial in skin infections and they had multiple minor changes every time. When I would incorporate their changes or when I would try and dispute their position, it would take months of haggling to get to a decision."

"Haven't you met with Europe or the FDA yet?"

"Sure. Several times each. They have essentially agreed with everything we've suggested as far as the trial design is concerned. They had more questions on other things."

"Like what?"

"Sorry. I can't get into that with you."

"Do you have enough money to get the trials going?"

"Yeah. Ullanger had to pay a large penalty for bailing out. We should be fine."

"Well – do you need any help from me or are you going to be OK?"

"I'm not sure that we'll be OK, but I don't think there is anything you can do that we're not already doing."

A few months later, Daniel saw that C-Biotech had started one of its trials in skin infection. They would need to complete two to get approval. Within a few months of that, the FDA announced at a Gordon Research Conference that they no longer knew how skin infection trials should be designed. Daniel was there and was in the forefront of discussing this problem with the FDA and with a number of his clients. Just after this announcement, Daniel got a call from Cal.

"OK. We need you again. I'll send over a confidentiality agreement and a contract."

After completing all the paperwork, Daniel headed out to Cleveland. It had been maybe two years since he had visited the last time. C-Biotech had enlarged their office space but the labs were smaller. It was a beehive of activity. Daniel and Cal sat in a conference room. Cal offered some coffee and then just looked Daniel in the eyes and sighed. Cal was completely bald, about Daniel's height, had brown eyes and sported a small goatee. His brow was furrowed and his eyes were sunken. Daniel thought that he looked like he needed sleep.

"You look like shit." Daniel said.

"That's how I feel. Let me close the door."

Frank's presence was felt even in his absence. The atmosphere was tense.

Cal said, "Well, we finally started our first pivotal trial. Enrollment is picking up quickly. We had planned to start a second one in a few months. We have all the drug supply we need and are just waiting on a few regulatory approvals. We have a contractor running the trials for us and have paid them 40% for both trials upfront and non-refundable."

Daniel quickly calculated that they must have already invested about $25 million or so just to do the trials. This didn't count any other expenses like manufacturing if that was still required.

Cal continued, "We don't know what we should do at this point. We don't know if the FDA will accept the design that they agreed to last year or not. We have pretty much decided to delay the start of the second trial if the FDA doesn't come out with something more definitive soon. But we don't know whether to continue the current trial or to bail before we get in too deeply."

"Have you spoken to the FDA?"

"Of course. Several times. They are completely unhelpful. They say they are re-evaluating their entire approach to antibiotic development starting with trials in skin infections. But they offered no insight as to whether they thought our design would be adequate or not even though they gave us their stamp of approval six months ago. They say that the 'regulatory science is changing' and that they have an obligation to keep up with these changes."

"That's BS. The science is not changing. We have learned nothing new as far as I know. It's the FDA that's changing and they're going to drag some companies into bankruptcy while they do it. That might include C-Biotech."

"OK. But what we would like from you is an opinion on whether to keep on going or stop the current trial now. I'll leave you with the investigator's brochure and the protocol that we are using and that the FDA and Europe both approved last year. I'll also show you the correspondence we've had with the FDA starting with our exchange last year and then since their announcement at the Gordon Conference in April. Let me know if you need anything else and let me know when you've finished going through everything."

Cal continued, "What time is your flight? If you need to stay longer than today – can you or do you need to get back?"

"Its OK, Cal. Relax. I'll get this done today and give you my thoughts on the whole thing."

Cal left Daniel in the conference room with reams of paper. Daniel started digging through the FDA correspondence first. Three hours later, he walked out of the conference room, helped himself to another cup of coffee and headed over to Cal's office.

"My opinion is that you can't win. I think that it is unlikely that the FDA will accept your current design – at least for now. If you keep going, you'll be sending good money after bad. But, if you stop, you will have to stop everything while you wait for the FDA to decide what they will do. I would guess that would take another year. Then you would have to try and restart pivotal trials, assuming that the FDA comes up with something that's actually feasible to do. Would you have enough money to absorb this?"

"Yeah. I'm thinking along the same lines you are. I'm not sure where Frank is on this. He might be so delusional as to believe that somehow we can salvage this trial. And no – we won't have enough money left to support two new pivotal trials even if we jettison our current trial now. We would have to raise more money."

"You won't be able to do that until the FDA provides a feasible path forward. What investor will throw his money down on FDA roulette? What will you do?" Daniel asked.

"Me? Who knows? You know, I really enjoyed practicing medicine before I joined industry. At Gorman, they let me keep working in a clinic once a week when I was in town. I did it gratis. I'm thinking of chucking the whole antibiotic development business altogether and just going back to seeing patients."

"Hmmmm. From what I hear from my friends in practice, that's not as much fun as it used to be either."

"Like I said. Who knows? One thing is sure. I won't hang around here to wait and see whether the FDA can get its act together and then to see if C-Biotech will be going bankrupt or not."

"I hear you. Good luck, buddy. Stay in touch whatever you do. Do you want me to write something up so you can give it to Frank? I can make it pretty detailed or give you a 100,000-foot view in one or two paragraphs. What would you like me to do?"

"Why don't you give us the two paragraph version. I'll let you know if I need more."

"OK. But promise that you'll at least let me know what C-Biotech decides to do."

"You got it."

Daniel sent in his report. C-Biotech kept their trial going another two months before they finally decided to pull the plug. Cal took a job back at Gorman having left there just two years earlier. But he would not work on antibiotics. He was going to work on drug development policy in a more general way. He would be paid well, have less aggravation and have to travel less. And he went back to his weekly clinic. He seemed happier when he and Daniel caught up over a beer during the next big infectious diseases meeting six months after Daniel's trip to Cleveland.

In the meantime, Henry Maloon worked with the FDA directly to establish a way for them to look at new trial designs for antibiotics. He got them to fund an effort by the Foundation for the National Institutes of Health to help. The FNIH put together a very large team composed of clinicians and scientists from a number of both large and small pharmaceutical companies as well as from the Infectious Diseases Society of America. Daniel was not invited to participate but rather followed their activities through his numerous contacts who were on the team. By the time the FNIH had made their proposal and it was accepted by the FDA, a year had passed. The design developed by the FDA was a feasible one – the trials could actually be successfully completed. But the endpoint for the trial that they chose, that the skin lesion should stop advancing within 2 days, was never widely accepted by clinicians and scientists nor by insurers who wanted to see patients cured and wanted that to be the endpoint for trials of new antibiotics.

C-Biotech never went bankrupt, but they were never able to raise any more money either. They shuttered their labs and fired most of their scientists as well as most of the rest of the staff. They hired a clinician from Penfrel who had been involved in the teracil project. But CL-219 could not progress. In spite of their new Chief Medical Officer, and in spite of feasible guidelines for carrying out trials in skin infection, C-Biotech was unable to raise enough money to fund two new trials.

Daniel thought that there were now two problems with CL-219 that might explain C-Biotech's conundrum. With all the delays, a number of new antibiotics that could be given in pill form and that would work on the MRSA superbug had progressed into late stage trials catching up with CL-219. Where CL-219 and C-Biotech had an advantage two years ago, they had lost it with the FDA perseveration.

The other problem was that with all of the delays over so many years, CL-219's patent life was starting to run out. The number of years of sales remaining before generic intrusion would be limited and therefore the value of the drug over its lifetime was diminished. In Daniel's opinion, there was nothing wrong with CL-219. A company that was unable to maintain its large pharma partnerships and that was ultimately undone by an incompetent FDA had dragged it through the mud. Daniel didn't think it would ever recover.

During these years, two biotechs did go bankrupt directly related to the FDA's inability to get its act together. While Daniel and most others assumed the worst, C-Biotech's new management was fighting hard. The ultimate fate of CL-219 and C-Biotech has not yet been determined.

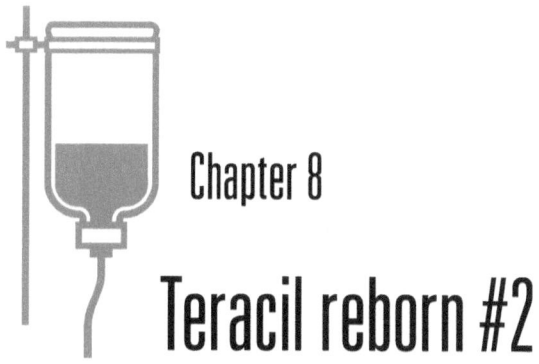

Chapter 8

Teracil reborn #2

Around the time that the FDA announced that they had figured out antibiotic trials in skin infections, Daniel got a call from yet another company, called Cycletech, that was also developing a teracil-like antibiotic. Its CEO was an old friend from Virnuc and he knew of Daniel's experience with teracil. Cycletech was located in Worcester, Massachusetts near the campus of the University of Massachusetts. The chemist whose technology formed the basis of Cycletech was a professor there. He had discovered a way to make teracil-like antibiotics completely synthetically, without relying on production of natural products from bacteria or fungi. Teracil and CL-219 were both still made by isolating such a natural product and then modifying it chemically. The Cycletech compounds were made entirely by their chemists in their own laboratories. Since Daniel still had a valid contract with C-Biotech, he had to terminate his contract with them in order to work with Cycletech. He notified both companies of this plan and both agreed. Frank at C-Biotech understood that he couldn't really prevent Daniel from working and C-Biotech was no longer using Daniel anyway. They parted on good terms.

The CEO of Cycletech was Marion Moore. Marion was a tall, thin, balding black man of around fifty years of age. He kept himself in good shape. Marion was the only black executive in either investment firms or biotech that Daniel had encountered. Marion's background was in marketing. He had worked previously for Gorman in their antibiotics marketing group before moving to Virnuc where he and Daniel worked together for a couple of years. Marion was well versed in the antibiotics market, but had little background in the actual science, or in the preclinical or clinical development of antibiotics. But Marion thought that he had something really good and he wanted Daniel's advice on how to take his products forward.

When Daniel arrived for his first visit, Cycletech's chief scientific Officer, Loren Greene, presented a 60-slide powerpoint overview of all of Cycletech's compounds in various stages of preclinical work. There was one lead compound that was just completing its first clinical trials – CY-309. There were several others, all further behind. Daniel admitted during their first set of meetings that he was confused. When he looked at the profile for CY-309, it was clear that, like teracil, this would be a very broad-spectrum drug – it would cover most types of pathogens both resistant and susceptible. The pathogens it seemed less likely to cover were those already resistant to teracil since, being similar molecules, teracil-resistant strains seemed resistant to CY-309. In fact, compared to teracil, there seemed to be little advantage for 309. Then when Daniel compared 309 to Cyceltech's other compounds, they were different only in that they covered a more narrow spectrum of pathogens. But for the bacteria that they covered, they seemed no better than 309 or teracil. When Daniel explained his confusion, Marion jumped in.

"I agree with your points. But first, what if we take 309 and direct our clinical development towards the resistant Gram-negative pathogens where there is more of an urgent medical need? Then, the other compounds could target more skin infections and respiratory infections caused by other bacteria. One of our compounds we thought could specifically be used as a pill and be used for resistant urinary tract infections caused by Gram negatives."

"OK. But you don't know yet that 309 cannot be used in pill form. Ideally, it would be available as an intravenous and as a pill form for use both in the hospital and outside the hospital. In fact, without something like that, I don't believe that 309 is different enough from teracil to allow you to take its market. But I do agree that if you target 309 development towards Gram negatives, you could target a different compound towards bacteria causing skin and respiratory infection."

"What if 309 had less nausea than teracil? Would that be enough to get physicians to switch?"

"I don't think so. The physicians I've spoken to do not seem very upset or worried about the nausea engendered by teracil. I think you need more. The best would be if you could make it into a pill. Teracil will never be able to be used in pill form. Plus there is a desperate medical need for an antibiotic pill that would kill resistant Gram-negatives. Think of all the hospital admissions we could avoid. Today, patients are being admitted to hospital just so they can get antibiotics administered intravenously and this is just because their infecting bacteria are resistant to all the antibiotic pills out there. What a waste of resources. Plus – hospitals kill people. Getting admitted to a hospital unnecessarily is not a good thing."

"OK, Daniel. Let us think about all this. I'll have to discuss this with the board as well. What else do you see that we need to think about?"

One area where Daniel was unsure had to do with how they elected to calculate the doses of 309 that were used in the early volunteer trials. First, they had elected to dose on a mg/kg basis – that is, they corrected for the patient's weight. The good thing about this was that it a avoided an important problem for teracil. Teracil was dosed at 50 mg twice per day. This low dose of teracil was used to minimize nausea but was not high enough to achieve therapeutic levels in certain very obese patients. This almost certainly caused some treatment failures in these patients. By dosing 309 according to the patient's weight, Cycletech was probably avoiding this pitfall. But, although pediatricians were very used to dosing by weight, most internists and physicians who cared for adult patients were not – so this would be a small commercial hurdle to overcome. So far, in dosing volunteers, they had exceeded their predicted therapeutic dose by 3 fold. At this highest dose, they encountered levels of nausea and vomiting similar to those seen with the marketed dose of teracil. At lower doses, where Cycletech thought that they would be therapeutic, there was almost no nausea and vomiting and the intravenous infusions were well tolerated.

Daniel noted that Cycletech had not used the standard mouse models of infection to calculate the therapeutic dose for humans. He raised this with Marion and Loren. They replied that they had not wanted to invest the money in those studies before getting the volunteer data. But they both agreed that with favorable data in the early trials, it might be time to invest in the formal animal models to be sure that the therapeutic dose they envision would be substantiated.

The other problem with the trials in volunteers so far was that they had not administered the 309 IV solution to the subjects orally as yet. This would be the first, quickest and simplest way to see if there was any potential to make 309 as a pill.

Daniel went on to point out that Cycletech was about to enter CY-309 into later stage clinical trials, but had no clinician on staff. They were unable to think clearly about a clinical development plan, plus they had no internal expertise in clinical trials. That would put them at the mercy of all the contract research organizations out there eager for business but who require supervision.

"You need to hire a CMO and you need a scientific advisory board with microbiologists and clinicians who are thought leaders."

Marion asked, "Would you organize that for us and be the Chairman of the Advisory Board?"

Daniel's first meeting with Cycletech made him start to rethink antibiotic marketing in some ways. Marion reminded him of a discussion Daniel had with Alistair Campbell during his days at Parabiotics. At Parabiotics, the company was suggesting making a number of different combinations of penicillin-like drugs with their B-lactamase inhibitor,

PB-100. When Daniel suggested that they might compete with each other on the marketplace because they had antibacterial activities that had some overlap, Alisatir just laughed. "Oh! What a terrible problem! I'll be selling three different antibiotics at three different prices to three different populations but each one might overlap to some extent with the other. I wish all my problems were that serious!" Daniel had to admit that both Alisatir and Marion had a point. By correctly designing the trials of the antibiotics – by studying them in different infections – you could successfully differentiate one from the other even if there was some overlap in the types of bacteria killed by the different drugs.

After those first meetings, Marion and Daniel both, separately, spent a good deal of time thinking about strategies for Cycletech. Marion started a search for a CMO and passed several CVs on to Daniel for his review. One of the possibles was Barb Mankowski. They had tempted her enough to come in for an interview at Cycletech. Daniel was thrilled with the idea that they might be able to work together again. But when they met at a bar in Worcester after Barb's interviews at the company, Daniel could see that this was not to be. Barb was tied down at her large pharma job with too many golden handcuffs. She wanted to wait for her bonus coming next spring. She wanted to see if she would get a promotion later in the Fall. Basically, Barb was back in big pharma mentally and emotionally and she was not about to make the move back to insecure biotech-land.

Marion kept bringing other candidates in and finally settled on one, Paul Herman. Daniel did not interview Paul, but Marion did send Daniel his CV. Paul was an infectious diseases physician and had been working in biotechs on the clinical development of antiviral drugs and vaccines for the last 10 years. Beyond what he knew from his days of treating patients, he knew next to nothing about the discovery and development of antibiotics in industry.

Daniel's initial impression was that this was going to be a long, hard slog. But when Paul and Daniel met two months later, Daniel was pleasantly surprised. He found Paul to be affable, smart, and most importantly, to be keenly aware of areas where he lacked knowledge. Paul kept himself in good shape. He was balding, about 45 years old, thin and about the same height as Daniel. Daniel thought that Paul would work hard to figure things out and would do so focused on his areas of weakness.

Daniel also found a number of areas of strength in Paul's experience and approach. He knew a number of the contract research organizations and had had a good deal of experience in managing these folks. That was going to be crucial as Cycletech advanced 309 into later stage clinical trials. He also could stand up for himself. Paul questioned the occasional BS thrown at him by Loren or even by Marion and even, occasionally, from Daniel himself. As time went on, Paul learned to say a forceful "No, we're not doing that!" Mostly, but not always, he got his way.

In subsequent meetings, Loren and Marion brought up the possibility of applying for government grant funding for several of their projects. Daniel was not enthusiastic. Virnuc had three large government grants when Daniel was there. Daniel himself was the Principal Investigator on two of them. These were two large research grants from the European Commission where the bureaucratic reporting obligations were onerous and where the company had to promise to manufacture the product (yet to be discovered) if any of the multiple academic partners wanted to take it forward into clinical trials. This would be true even if the company had no interest in the product at that point.

The third grant at Virnuc was a large one from the National Institutes of Health in the US to support early trials of one of Virnuc's Hepatitis B drugs. Those trials were never even started because the NIH could never identify a practical trial design that would allow their sites to actually enroll patients. Daniel's experience with these grants at Virnuc and the amount of useless man-hours of work dedicated to supporting these grants made him very skeptical. He shared his war stories with Marion and Loren. Marion remembered these grants from his time at Virnuc, but he was determined. "This is a great way to get non-dilutive funding for our projects outside of 309. And, to be honest, our investors are having a hard time justifying us spending time on these other projects even though we think we could have valuable products out of them some day."

"Yes. I understand. There are lots of companies applying for similar grants for the same reasons as you. But just be prepared for the opportunity cost you will have to bear."

Loren chimed in "But these grants all pay for administrative overhead. What opportunity cost?"

"The one involved when you actually have to supervise all those administrators and sign off on what they are doing." Daniel retorted.

Marion and Loren then outlined their plans. They wanted to get the government to support development of one of their early, preclinical compounds that they were targeting for the treatment of respiratory and skin infections. But the drug was active against certain so-called special pathogens. These are pathogens like *Bacillus anthracis* the agent of anthrax and *Francisella tularensis* the cause of tularemia – where both could potentially be used for biological attacks. They also wanted government support for the 309 program based on a similar rationale.

"Will the government actually fund both programs given the similarity of the compounds?"

"Apparently they will since they are willing to hedge their bets in case one project should fail at some point along the way."

"And," Daniel added, "you will have to deal with the added bureaucracy of dealing with bioterror agents – no small encumbrance."

Marion and Loren persevered and Loren, through 70-hour workweeks was able to come up with several million dollars in funding for their early, preclinical project, but more importantly, she landed a 70 million dollar grant for the 309 project. These monies allowed Cycletech to pursue additional trials for 309 that they might not otherwise have carried out, and it allowed them to keep a robust antibiotic discovery activity alive at least for a few more years. Yes, there was an opportunity cost, but the scientists and Cycletech all thought it was worth it to keep their early programs going and to increase the potential value of 309.

While all this was going on, Paul was busy finishing the early volunteer trials for 309. In these trials, following Daniel's advice, he included a small arm where he administered the IV solution of 309 to some of the volunteers by having them drink it. He wanted to see if there was any potential to have a 309 pill. Daniel was adamant that they try this because he thought that without a pill form, competition with teracil, since it was already on the market, was going to be difficult. The data suggested that a pill of between 400 and 600 mg would be required to give levels equal to the weight-based dosing Cycletech had chosen for the IV form of the drug. There was hope for the pill.

Loren, in the meantime, had gotten busy getting 309 studied in the standard animal models of infection that would allow an accurate prediction of the human dose. But here, Cycletech ran into a roadblock. 309 was not working in these models. They tried the same bacteria that Penfrel had used for teracil where teracil worked reasonably well. But even at high doses, the antibacterial effects in the mouse models were minimal. Daniel went back to review the data that Cycletech had amassed in other, non-standard models. Here, 309 worked. But the data was not detailed enough to allow a truly accurate prediction of the human dose. The contractor who had carried out the studies for Cycletech wanted to pursue a number of different dosing regimens and to carry out the studies for a longer period of time. But Daniel argued that although this might make sense scientifically, the proposed experiments were not validated and it was not clear that they could be readily used for a calculation of the human dose as required by the regulatory agencies.

"So – now what?" Loren asked.

"So, we go with what we have." Daniel replied. "The doses you used in the non-standard models were still similar or lower than the actual human doses you have targeted in your volunteer studies. We're just going to have to say that for this drug, the standard models do not work well. It won't be the first time the FDA has seen this problem and it won't be the last. Then we'll use our human data from trials in patients with infections to justify the dose we choose for the final trials and for the marketed drug. I don't see any other way to do this."

There was a lot of head nodding going on around the table. But no one was sure how the FDA would react. Daniel suggested that they get a protocol and an investigator's brochure together for their first trial in patients and get them to the FDA as soon as possible in order to understand whether there were going to be problems with their dose justification section.

"But we haven't even decided what infection we should study for this first trial." Paul said.

"Complicated intraabdominal infections." Daniel replied. "This would include appendicitis where the appendix has already started to rupture into the abdomen, infections of the gallbladder, infections of the pancreas, perforations of the stomach or intestine where infection has already started and abscesses in the abdomen. These trials enroll rapidly, are very straightforward with designs available from many other trials and you get good microbiology. In fact, in the intraabdominal infection trials we ran at Parabiotics, we enrolled the entire trial in less than a year and had 25% of our infecting organisms as resistant Gram-negative superbug pathogens infecting the patients we studied. That was a great trial. Let's do the same thing."

Daniel also pointed out that without a strong dose rationale, they would probably need to study at least two different doses. The problem with this is that you can't really give infected patients a lower dose than one you think will be effective. You can skirt that line to some extent – but that's all. Paul and Daniel picked two dosage regimes that were fairly similar. One was a once per day dose where the amount given was higher. The second was a twice per day dose at a lower dose each time – but where the total daily dose was greater. If 309 was similar to teracil, it would not matter how many doses were given per day. This allowed them to test this in a limited way in patients and at the same time, two different total daily doses were also being tested.

Paul took Daniel aside later saying that he would need help with this one since he had never done one of these before. "OK. No problem. We'll get this done. Who will handle the microbiology part – you or Loren?"

"Loren."

Daniel called Loren and discussed what would be necessary from her to support a trial. At a minimum, it would mean establishing a central microbiology laboratory where all organisms would be sent by the participating hospitals. The central lab would identify all the organisms and determine their susceptibilities to a variety of antibiotics including 309 and its comparator whatever that would be. Daniel provided her with four different contractors he had used for this in the past.

The next several months saw frenzied exchanges of documents between Daniel and Paul and Loren as they tried to get everything lined up to send to the FDA and to actually get a clinical trial off the ground. Loren put out a requested for applications for a central microbiology lab to three of the companies recommended by Daniel. Loren wanted to use a European company she had used before that was on Daniel's list as well. But they came in with the highest bid. "They always do this." Daniel said. "I don't know if they just try their highest bid first hoping that you will fall for it or if everyone else has one set of prices for big companies and another for biotech. Whatever the explanation, if you just tell them to try again and to bring their bid down by 25%, we should be OK." They decreased their bid by 20% and Loren engaged them.

None of the clinical research organizations Daniel had used before responded to Paul's request. Daniel assumed that they were tied up with doing late stage trials in intraabdominal infections for other companies and could not take on the extra work. So Paul and Daniel had to go through the other companies that neither had worked with previously. Daniel made a few calls to colleagues and was able to help Paul narrow the selection down to two that had come highly recommended by others. "Remember, Paul, no matter which CRO you pick, they will always be a pain in one way or the other. You will have to watch them closely no matter who you pick." Daniel warned.

Finally, three months after their last meeting, Cycletech sent their documents to FDA. One month later, they received comments back from the agency. To Daniel's relief, there were no questions on their dose justification. The FDA seemed at least willing to see the data from this first trial in patients before deciding further on the dosage regimes to be studied. There were no questions on the dosing regime. There were a few questions on the plan to study 309 toxicity and the FDA requested more data on the activity of 309 against more Gram-negative pathogens. But all in all, the FDA seemed favorably disposed to Cycletech's plans for 309.

Paul had already engaged a contractor to carry out the clinical study. The contractor had proposed sites in the US, India, Lebanon, Ukraine and Estonia. Paul had already visited all the sites and agreed with the contractor that they would start. Both Paul and Daniel realized that the sites might have to change. Not all would live up to their promise of delivering patients and others may have other problems prompting the cancellation of some sites and the addition of others. But the first patients were enrolled in the beginning of 2011.

As the trial was being conducted, Cycletech received news on the grants for which it had applied. Both were going to be funded for a total of something like $70 million. Loren was the point person at Cycletech for both the grant for 309 and for the one supporting Cycletech's earlier programs. She was rapidly overwhelmed with work just keeping up with the reporting requirements to support both grants and with the unending meetings with government contract supervisors. Daniel could see that Loren's work on microbiology support both for early programs and for the support of 309 was suffering. At the same time, he knew she was working more than her usual 70 hours per week since he was getting emails from her at all hours of the night and on weekends.

One year later, the trial was completed and within a few months after that, the data were analyzed. The results clearly indicated that 309 was at least as good as the last line and gold standard antibiotic for intraabdominal infections that was used as the comparator. In the trial, just as in the Parabiotics trials, about 25% of patients had highly resistant Gram-negative superbug infections and 309 worked well for those patients. Both of the doses of 309 were equally good and neither had big differences in side effects. The Scientific Advisory Board for Cycletech met and confirmed that they agreed with this assessment. They also noted that if an oral form were available, the drug would be a very important addition to the antibiotic armamentarium of clinicians and their patients.

Paul and Daniel sat down and rolled up their sleeves to prepare for the final trials that would be required to get approval from the FDA and European regulatory authorities. At this point, the regulatory authorities both in Europe and the US said that a single trial in intraabdominal infection and a single trial in urinary tract infection would suffice to justify approval for the treatment of both types of infection. This was new and cut the number of patients required for studies in half. Paul and Daniel decided to simply repeat the phase 2 intraabdominal trial as a larger phase 3 trial. Then they set about designing a phase 3 trial in urinary tract infection.

For the UTI trial, there were some interesting challenges. The most widely used drug that came in both IV and pill forms, was already generic and was sold at a very low price. In Europe, the price you can negotiate for a new drug with the various national authorities depends, in part, on the price of the comparator used in your trials. Daniel wanted to get around this problem by showing in the trial that 309 would actually be superior to the generic antibiotic because of infections resistant to the generic.

Paul balked. "No one is going to allow you to randomize patients to possibly getting an antibiotic where they already know the pathogen is resistant. Its not ethical."

"Of course not – and we won't do that. When most patients are entered into a clinical trial, the pathogen is unknown. What we'll do is first exclude patients who are very seriously ill with their infection to make sure we don't hurt anyone in the trial. Then, we'll ask the clinicians to ignore the microbiology report on their patient and decide what to do based on how the patient is doing clinically. If they are recovering, they continue on their antibiotic whichever it is (and the clinicians are all blinded in terms of the therapy anyway). Then, if the patient is not doing well, we will allow them to either take the patient off study and treat with another antibiotic, or to allow them to switch blindly to the other arm of the trial – not knowing which antibiotic the patient is getting nor which they will get after the switch. In either case, that particular incident is a failure for the initial antibiotic. We can then compare both types of failures – those taken off study – and switches between arms – as an indication of which drug is more likely to succeed."

"Sounds like a statistical nightmare to me." Paul said with raised eyebrows.

They agreed to bring on yet another clinical trial expert as a consultant as well as a statistician to help them. In the end, they constructed a trial where there would be a stepwise analysis. First, the data would have to show that at least 309 was statistically no worse than the comparator based on FDA guidelines. If that were true, then an analysis of those patients who happened to have infections where the generic antibiotic's activity was at least questionable would be undertaken.

The FDA was unhappy. "This is just a subpopulation analysis. There will not be sufficient numbers for us to use this and we think that in spite of your precautions, patients might be placed in a dangerous spot." they said. Paul argued that since the generic antibiotic was already the top choice of physicians for UTI anyway, the patients would actually be better off in the trial since they would be likely to get 309 that would be more likely to work. He also presented them the statistical plan showing that their projected numbers would provide the potential for statistical superiority at the end of the trial. The FDA remained skeptical, but the European authorities were enthusiastic about the design.

Paul and Daniel's UTI trial design won the day.

Chapter 9

The One Bug Wonder

Daniel received an email from another European company – this one in Barcelona – that was developing a very peculiar antibiotic. It was so unusual that his interest in it was instantly piqued. The antibiotic they had worked only against a single bacterium – *Acinetobacter*. *Acinetobacter* is one of the most fearsome of the Gram-negative superbugs because it is so resistant to antibiotics. Daniel participated in a study in the US showing that 80% of these isolates from patients were resistant even to our last line antibiotics, the carbapenems. The only antibiotics available to treat these organisms were either toxic or had never been shown to work in the most serious bacterial infections. So a new antibiotic that would be active against *Acinetobacter* was sorely needed. The problem was that there are not that many patients infected with this bacterium. In order to make money on such an antibiotic, the company would have to charge a very high price. Also, it was not clear at all how you could carry out clinical trials to demonstrate that the antibiotic worked. How many patients could you enroll? What kind of infections would they have?

All of these questions prompted Pepcore to contact Daniel. Daniel responded with enthusiasm by immediately calling Jose Diaz, their Chief Medical Officer. Jose was surprised to get a phone call almost immediately after having sent an email message. But he started explaining his company and their drug to Daniel. One family owned the company that started out in the chemical business. They evolved to the current company, Pepcore, to explore new molecules based on the building blocks of proteins, the amino acids. These molecules are called peptides. They would identify active peptides, then modify them to be more like non-peptide compounds. They had to do this because most peptides are not suitable as drugs – the body degrades them to their amino acids very quickly. Pepcore was exploring a number of these molecules in various therapeutic areas – heart disease, viral infections and antibiotics. Their anti-*Acinetobacter* drug had just finished its first trials in human volunteers, and Pepcore was not sure how to go beyond this stage. Jose agreed to send Daniel Pepcore's usual contract agreement.

A month later Daniel was in Barcelona for his first meeting with the Pepcore scientists and Jose. They spent an entire day going over the data they had amassed on their compound, called Pep-2800. They started with a peptide that had activity against a broad range of bacteria, but that also killed human cells. Gradually, they were able to improve the activity of the peptide against certain types of bacteria while losing the toxic effect on human cells. Finally, as they did their chemistry magic to turn the peptide into a drug, they only had activity left against one type of bacteria, *Acinetobacter*. But it was very potent activity. There was no longer any toxic effects on human cells at any concentration they could test.

Pepcore had established a collaboration with a group of university scientists in Madrid who were able to identify the way that Pep-2800 killed *Acinetobacter*. It inhibited the synthesis of the bacterium's cell membrane. Without the ability to make its cell membrane, it could not divide and it died. The enzyme targeted by Pep-2800 had never been targeted by any antibiotic – ever. Pep-2800 had no other activity. This meant that it would kill *Acinetobacter* resistant to any other antibiotic and that if you could get *Acinetobacter* resistant to Pep-2800, it would not be resistant to any other antibiotic. In fact, the Pepcore scientists and their academic collaborators had a good deal of very convincing evidence that this was all true. Daniel, on seeing all the data, was even more excited than Jose and the Pepcore scientists who had been working on this project for the previous five years.

There were a number of problems that Daniel could see immediately. Pepcore had never carried out the key animal studies required to determine human dosing. They did have animal studies where they showed that Pep-2800 would work against pneumonia and deep soft tissue infections – but those studies were carried out at a fairly high dose. Daniel also was surprised to see how small their first human volunteer studies were. For each dose, only four volunteers received active drug and one received a saline placebo. These very small numbers made it difficult to see whether there were truly side effects from the drug or not. He also saw that Pepcore only dosed the volunteers for five days when he knew that therapy of serious infections was likely to require at least ten days of dosing. When Daniel raised his questions at the end of their first day of meetings, Jose looked at the floor and mumbled a response.

"What? I didn't get that."

Jose lifted his eyes to Daniel and said in a louder voice, "Our budget was too small to allow the animal studies you suggest nor did we have enough money to carry out a bigger and longer volunteer trial."

Daniel tried to find out why their budget was so tight, but Jose would not say anymore. Daniel agreed to send a written report on his evaluation of the data presented during the daylong meeting along with suggestions for next steps. He thought that studying just two or maybe three doses in a larger number of volunteers for a longer period of time would be wise. After he submitted his report and his recommendations, he heard nothing from Jose for several months. This stuttering communication would be typical of Pepcore's relationship with Daniel. When Daniel did hear back from Jose, it was to request another meeting to review progress on 2800. Daniel suggested that they provide him with an agenda and with the data they wanted to review ahead of time. They did. It arrived by email the morning he arrived in Barcelona for the scheduled meeting.

Jose started the meeting by reviewing the final data from the volunteer study. Pepcore had decided that they would not expand the volunteer study as Daniel had suggested, but would take the risk of going to longer therapy in infected patients as their next step. When Daniel asked about how this would be received by the regulatory authorities, Jose assured him that it would not be a problem. Daniel tried hard not to raise his eyebrows. "OK."

Then Jose started a discussion on a design for their first trial of 2800 in infected patients. Pepcore was trying to design a so-called non-inferiority trial in patients with very serious pneumonia caused by *Acinetobacter*. They would compare Pep-2800 to the best antibiotic available for these infections, a carbapenem. The problem for Daniel was that so many of the strains would be resistant to the carbapenem that they would not be able to recruit patients. The other problem was that physicians, not having confidence that 2800 would work, might be reluctant to enroll their patients in the trial where some patients would receive only 2800.

Daniel had an entirely different approach. First, he said, Pepcore would have to carry out the animal studies he had suggested at the last meeting several months ago. Jose interrupted him to point out that those studies were in progress and that results would be available in another 1-2 months. But Jose wanted to know why these studies would be so important for Daniel. "Because they would give physicians confidence that the drug at the doses being proposed for the trial would actually work in these very ill patients. Without some level of confidence that the drug will work, it will, rightly, be impossible to enroll patients in the trial." Daniel wanted to know who was doing the animal studies, which infections would be studied and how many strains of *Acinetobacter* would be examined. Most of the rest of the day was spent discussing the experimental design. Daniel was still not satisfied that the study would be robust enough to convince physicians to enroll patients. At the end of the day of meetings, Jose asked Daniel to explain what kind of trial he envisioned for Pep-2800.

"I'm thinking about a very small trial. We would enroll only a few hundred patients. They could have either pneumonia or soft tissue infection. I would also suggest urinary tract infection, but it looks like 2800 does not achieve high enough concentrations in the urine to work there. There would be no patients enrolled in a control arm. Our controls would have to be derived using some sort of historical data. We can discuss how we would obtain these data later. Then we would treat all these patients with a combination of Pep-2800 and a carbapenem. But the only patients we would examine to decide if Pep-2800 was working or not would be those infected with carbapenem-resistant pathogens.

The reason we need physicians to have confidence in Pep-2800 is that for those patients 2800 will be the only active drug on board. If this works, you could be approved and you could market your drug just based on this single small trial." Jose pointed out that physicians liked to use combination therapy for these patients. Daniel agreed and suggested that they cold add an additional antibiotic – but only for the first three days. Otherwise the additional antibiotic would contaminate the trial. This would continue to be a point of discussion but Daniel said that there was no point in doing anything further until they had the data from the animal studies. Their meeting ended on that note.

Daniel, Jose and a growing team of Pepcore clinicians and microbiologists would continue to meet for the next two years. Pepcore did obtain animal data. Daniel managed to convince them to do more work to get a really strong data package together that would be convincing to physicians and the regulatory authorities. Pepcore followed his advice on this one.

The Pepcore team thought Daniel's plan for a first trial in infected patients was too risky. They wanted to find a way to develop Pep-2800 with less risk. Pepcore finally settled on a design where they would compare Pep-2800 added to a standard antibiotic compared to treatment with the same standard combination plus an antibiotic called amikacin. In this way, Pep-2800 would be compared directly with amikacin as a standard add-on type therapy.

Daniel disagreed strongly with this approach. He thought that, at best, Pepcore would show that Pep-2800 was as good as or at least no worse than amikacin. But for pneumonia in particular, amikacin was known not to work well at least when given alone. Daniel said, "Your trial will show that standard of care is similar to standard of care as far as physicians are concerned. You will not be highlighting the activity of your drug against resistant infections in your proposed trial at all." This disagreement would continue until Daniel finally retired seven years after having started work with Pepcore. Daniel frequently asked the Pepcore team why they continued to consult him since they clearly disagreed with his recommendations. He never did understand that.

Chapter 10

Gorman Ltd.

Gorman Ltd first called Daniel just as he began consulting. They even completed a consulting agreement. But in spite of the fact that Daniel knew many of the scientists at Gorman for many years, they did not actually use him as a consultant until just a couple of years before he retired. And he did not understand why they suddenly called on him at such a late date. But he was happy to get together with so many old friends and colleagues and work on their antibiotic programs. Their meetings always took place at their facility outside of Philadelphia. Daniel remembered the site from his meetings with them when he was at Penfrel and was trying to pry their Beta-lactamase inhibitor from them. Even though Gorman held a majority interest in Virnuc, Daniel never attended meetings with their antibiotics group while he was working at Virnuc.

At their first meeting, Gorman's scientists and clinicians gathered outside Philadelphia along with Daniel and several academic advisors. A number of different antibiotic research programs going from those still in the test tube phase to those already being studied in infected patients were presented. Daniel felt rather lost. Gorman's presentations were designed around specific questions that they wanted to pose to their advisors. But Daniel wanted to have more depth and context for each project and felt uncomfortable trying to answer their questions based on limited sets of data. He also felt alone since the other advisors seemed happy to provide the kinds of answers Gorman was looking for. He grew quiet during much of the meeting and he knew that his silence was noticed. Quiet was unusual for Daniel. During a break one of the scientists Daniel knew from way back in the Smith-Kline Beecham days at Brockham Park, Bert Mankin, approached him.

"Are you OK? You are quiet today."

Bert was one of the few survivors of Brockham Park to come to the US to work on SKB's antibiotics programs. He later, like many of their scientists, ended up at Gorman whose labs were not far from the SKB site. Bert had married an American and raised his children in the Philadelphia area. He was short with blond curly hair and retained a strong, Manchester accent.

"I don't feel comfortable without seeing more data, Daniel replied. The context is missing for me. Also, I don't understand why you are pursuing a number of these projects – but I know that this has got to be my problem and that you must have data that persuades you that you're on the right track."

They agreed to have a more private chat later. That occurred after the traditional advisory board dinner the night after the meeting. Daniel and Bert left the rest of the group at the end of the meal and headed to the hotel bar. Daniel explained that in order for him to help Gorman and to provide answers to the questions they were posing, he would need to see a complete data package for each of the projects. He asked Bert if the other advisors had seen the data perhaps at some previous meeting. Bert was evasive, but promised to send Daniel more data on each project.

Back at his home office in Connecticut Daniel received a series of emails from Bert each with a large, password-protected file. The passwords were all sent in separate emails. The list of projects included:

A drug with a structure related to the quinolone antibiotics (like Cipro) called GM-908. It could only be given intravenously and was active against staph including the MRSA superbug. Immediately, Daniel worried about all the toxicities that are associated with the quinolone antibiotics. They were toxic to cartilage – especially in children. They caused tendon ruptures in adults. They could be toxic to the heart, they could cause high blood sugar or low blood pressure. About 40% of the quinolones that had been studied in humans in advanced trials or had made it all the way to the marketplace had failed or been withdrawn from the market because of toxicity. While GM-908 was not really a quinolone, it was close enough that Daniel was concerned.

In addition, Daniel thought that there were lots of other intravenous drugs around that treat MRSA infections. Why do we need another one? There was not a lot of resistance among MRSA to these other intravenous antibiotics.

GM-417 was another drug that could only be given intravenously. It was active against staph and MRSA and against pathogens that cause pneumonia – so it was more broad spectrum than GM-908. But at the same time, its target in bacteria was an enzyme that worked by using a zinc ion. GM-417 was shaped to fit into the enzyme like a key in a lock and to bind the zinc ion or knock it out of place. The problem was that many human enzymes used zinc or sometimes magnesium and sometimes iron as well. What if GM-417 could bind all of these metal ions? If GM-417 bound the iron that was in hemoglobin, for example, it would be severely toxic. In addition, based on the animal studies carried out by Gorman, the dose of GM-417 had to be several grams per day in order to effectively treat infections in people.

After going through all the data on all Gorman's projects, Daniel was more confused than ever. He only found one project that he thought was potentially promising and that one involved a pencillin-like drug that worked against Gram-negative superbugs resistant to almost everything else. That drug, GM-110, did not come from Gorman's scientists, but had been licensed in from a Korean pharmaceutical company. It was still in its early stage trials.

Daniel called Bert.

"Thanks for sending all the data. I finally understand the science behind your projects. But I'm more confused than ever. There are many other antibiotics that are better and less risky than your most advanced compounds, GM-980 and GM-417. Why are you so stuck on these molecules?"

Bert hesitated on the phone. "Because they both hit novel targets in the bacteria, there is no bacterial resistance to them yet and bacteria resistant to existing antibiotics are killed by both of these drugs."

"But who needs another IV-only Gram-positive-only antibiotic like 908? There are at least five or six of these on the market and there is little or no resistance to them yet. Plus – it might carry the same safety risk as a quinolone even though it hasn't shown up in your safety testing yet." This was met by silence. Daniel continued, "And GM-417 has to be given in massive doses and is a metal ion binder. Again, there is a toxicity risk here that is not present in a number of molecules ahead of 417 in development that you could license in. Why wouldn't you do that instead of continuing with 417."

"Ok. Listen. We had to fight hard with our management to let us proceed with these molecules. We pushed the idea of novel targets hard and they bought into that idea. Its going to be hard to go back and say that we want to in-license something else unless one or both of these start to fall apart."

"But if they do run into serious problems, how will that look for you and your team?"

"Not good."

So Daniel thought he understood. Bert and the Gorman team had pushed hard for molecules discovered by their own scientists in spite of potential shortcomings and risks. It was the "its invented here" syndrome. Gorman would rather spend money on home grown products than take a chance on someone else's molecules even though they might in fact be farther ahead in development and less risky. Daniel just shook his head. "OK. I get it. Good luck on all this. What should I do? Do you want a report or what?"

"That's OK. We'll just check in with you at the next advisory board meeting. Thanks for getting into all of this for us."

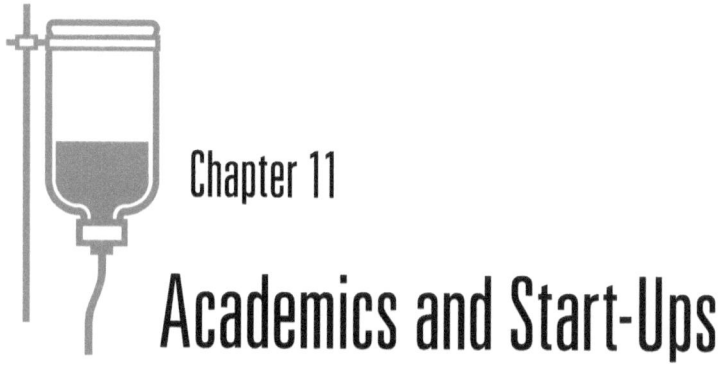

Chapter 11

Academics and Start-Ups

Daniel was, for a number of years, a member of a grant review committee for the National Institutes of Health where grant requests from professors were reviewed to decide whether they should be funded or not. This particular committee focused on research on antibiotics. During his years consulting, Daniel was also frequently called by university groups and by start-up companies usually founded by university professors trying to discover or develop new antibiotics. Based on these interactions, Daniel was firmly convinced that, with a very few exceptions, university professors, while being sometimes brilliant scientists, did not know the first thing about drug discovery or development. While Daniel always tried to be tactful and respectful of the scientists for whom he was asked to provide advice, he was scrupulously honest. This sometimes led to lost business.

One of the first start-ups to call on Daniel was located in Hartford, Connecticut, not far from Daniel's home and office. The company sold laboratory reagents – mainly antibodies – that scientists could use in their experiments. During their work to produce these antibodies for research, they came across some that they thought might be good for treating infections. After examining the little data they had, Daniel provided a review of the history of antibodies for the treatment of infections – there were none. None had ever worked. While he thought further research was needed and would be interesting, he suggested that the field was too high risk for this company to embark upon and that there were probably better ways to invest their excess cash. Apparently they agreed and never called him back.

Another start-up that consulted with Daniel was located just outside New York — a short train ride away on Amtrak. A scientist from the University of Pennsylvania who had discovered a way to make proteins into drugs formed the company. In this case, the drugs the company worked on were antibiotics that killed the MRSA superbug. When they first called Daniel, they had data on how well the drug killed bacteria in a test tube and they had information on the toxicity of the drug in rats. They had carried out some animal models of infection showing that the drug could work as an antibiotic in mice, but they did not carry out the animal experiments required to determine the required dose in humans as yet. As Daniel looked at the information supplied by the company before his first trip to visit them, he was struck by the fact that in the toxicity studies, rats suffered neurological toxicity at fairly low doses of the drug — not much higher than the doses showing cure in the animal infections studied so far. He also realized quickly that the drug could only ever be administer intravenously — it would never be a pill or capsule. Based on these observations, he concluded that the company should not proceed to develop the drug. Another IV-only anti-MRSA drug was not where there was a medical need. Lots of those drugs were already on the market or were close to the market — way ahead of this product. Also, he thought that the risk of encountering toxicity in humans was high. After reviewing all their data with the team, he recommended that they not proceed with this project suggesting that the risk was just too high that there would be no interest from potential partners.

This was a small company with one advanced project — the MRSA drug. They were completely dependent on the prospect of finding a partner to continue the development of their drug. They never called Daniel back — and continued through phase 1 trials in volunteers. They ran into the toxicity that Daniel had predicted. In spite of that, they were able to scrape and borrow enough money and use low enough doses of their drug to start studies in patients with skin infections. But none of this was enough to attract a partner with money and the company went bankrupt.

In his role as a reviewer of grants for the National Institutes of Health, Daniel got a good overview of the state of antibiotics research within American universities and medical schools. It was depressingly dismal.

Back in the 1980s Daniel and a number of colleagues complained that the NIH was not funding antibiotic-resistance research. They carried out a study of publicly available information from NIH showing at least a 30-year neglect of antibiotics research at NIH. Things did not turn around there until 2006. The legacy of their neglect was an almost complete absence of trained investigators in the field.

Daniel was extremely discouraged about the ability of researchers to even enter the area of antibiotic discovery and preclinical development. He spent some time visiting academic centers and speaking with colleagues in academia on anti-infective research. Although he was incredibly impressed with the level of motivation and the innovative approaches being taken in academia, he was equally impressed with the lack of real ability to understand what one can reasonably pursue and how to pursue it either preclinically or in the clinic. Daniel found that academic departments of microbiology in the United States, for example, lacked clinicians and therefore had no idea how their ideas for antibiotics might ever be applied to therapy for patients. They also lacked any understanding of how to begin to prepare a drug to be taken forward into human studies – the animal infection models required, the toxicology studies required, the quality of manufactured compound required. All of this was beyond their ken. Among those in academia today who actually have the skills necessary to discover new antibiotics and bring them forward, were those rare refugees from the shutdown of antibiotic research in industry who ended up working in a university somewhere.

Of course, the other problem Daniel struggled with was the fact that the NIH still put very little money into antibiotics research. During the years he was a grant reviewer, the payline for funding was never above 10% and usually hovered around 6-8%. That meant that of all the grants these committees reviewed, only a tiny number would actually be funded. This might be true even though there were more that might have deserved funding. This funding was also divided between antibiotics for bacterial infections, anti-fungal drugs and anti-parasitic projects like for treatment of malaria. In one session, the 30 reviewers present reviewed over 100 grants in all these different areas of research. At the end of it all, only seven grants were funded and these only included two for antibacterial antibiotic research. Daniel's conclusion from that experience was that the NIH was not likely to be the answer to the continued abandonment of antibiotic research by large pharmaceutical companies – at least not without more money being poured into the effort.

Based on all this, Daniel suggested to the NIH that they first invest in training academic scientists in the area of antibiotic discovery and development and that they do this by offering grants for training academics in the pharmaceutical industry. There are still a few companies active in antibiotic R&D who would welcome such academic trainees, he thought. After all the years of NIH neglect, they certainly owe the American public, the world and antibiotics investigators this much. But all his cajoling and arguments both in public and in private with the leadership of NIH were to no avail. The state of sophistication of grants being submitted by academics for funding of antibiotic research remains dismal. No training grants of the kind envisioned by Daniel are yet available through NIH.

One day, Daniel got a call from a professor of chemistry in Scotland, George Macallan. He was working on a project involving finding new drugs for MRSA superbug infections and he had submitted a grant to the Scotland Trust for over $25 million dollars. George had identified a way to get drugs to anchor to the bacterial membrane and to physically deliver the antibiotic to a place where it could be more active. While Daniel never thought that the plan to develop this for use against MRSA was going to work commercially given the number of MRSA antibiotics available or in late stage development, he thought that the scientific approach was a good one. He also thought that George was one of the few academics he dealt with who had a good understanding of antibiotic discovery. He started working with George to make sure he would get the funding.

Daniel viewed this as more of a demonstration project to show that George's scientific approach to getting antibiotics to deliver their activity to the place where the antibiotic targets were – close to the membrane – would work. Daniel said so at the grant review committee where George and his team presented to the Scotland Trust. The

Scotland Trust indicated by their questions that they were very interested in the possibilities for making any antibiotic that might emerge from this project a commercially valuable asset. Daniel tried to steer the conversation back to making this more of a scientific proof of concept type study rather than a commercially viable antibiotic.

One month later, George called to say that they had received their funding. The next five years involved quarterly meetings where Daniel would provide advice as to how to further optimize molecules and how to appropriately test them to show that they could be effective antibiotics in humans – but he never lost sight of the fact that this was, for him, still a proof-of-concept project. The Scotland Trust never lost sight of their view that this should be commercially viable.

Towards the end of the five years of grant funding, George and his team had a reasonable lead molecule that could be ready to go through the studies required to prepare for eventual clinical trials. But Daniel did not think the investment in those efforts would be worthwhile since the compound was not likely to be attractive enough commercially to bring in the funding required to actually start clinical trials. At this point, the Scotland Trust sent Daniel a long questionnaire on the commercial importance of George's lead compound and followed this up with an interview. They apparently ignored much of what Daniel said and proceeded to try and find partners – with no success. George, on the other hand, heard Daniel's advice and started another project using the same technology to hit Gram-negative superbugs where the medical need was much higher. But the scientific challenge was also much higher and after the end of the funding period, Daniel heard nothing more from George.

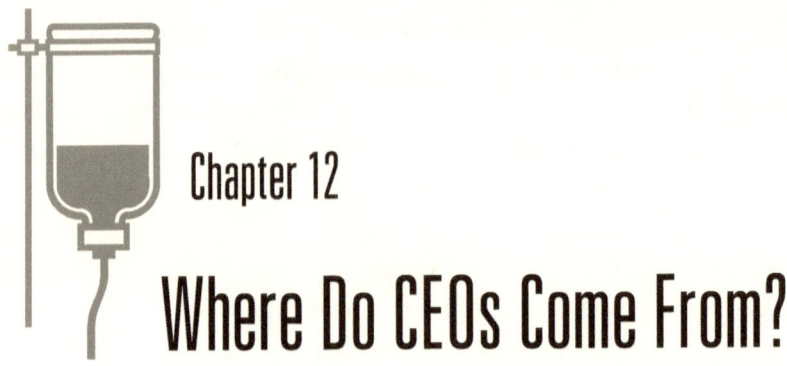

Chapter 12

Where Do CEOs Come From?

During his career in industry, Daniel ran across many CEOs. The companies he encountered ranged from those with four employees to those with 60,000. CEOs probably get more credit, and more blame, then they deserve. Its true that the buck has to stop somewhere – but the fact is that, given the complexity of the pharmaceutical business, there is no way a CEO can master all that he or she surveys. CEOs can be finance people, accountants, marketers, lawyers and deal-makers (business development types). They are only very rarely scientists and virtually never physicians.

Before Daniel even arrived at Penfrel, he already had a favorite pharmaceutical company, Merck, that came with his favorite CEO, Roy Vagelos Vagelos, the CEO and Chairman of Merck at the time was a physician scientist like Daniel. Vagelos had been at the National Institutes of Health in the 1960s where he conducted research on fatty acid synthesis. From there he went to Merck Sharpe and Dohme where he became head of Research and Development and eventually CEO and Chairman. During his tenure at Merck from 1975 until 1994, he led the company through its golden era of growth with a number of major products. One of them, Ivermectin, developed to prevent and treat parasitic disease in animals, was found to be active against the agent causing river blindness in people. Vagelos made the drug available free of charge for the treatment and prevention of that terrible disease and helped the World Health Organization distribute the drug to inaccessible areas of the world. Vagelos was Daniel's idol among pharmaceutical executives.

The first CEO Daniel encountered personally was Barry Boswell at Penfrel. He was a corporate lawyer who inherited Penfrel as a company that made everything from pots and pans and men's aftershave to pharmaceuticals. Barry was stern and ruthless. When Daniel first arrived at Penfrel, perhaps during his first six months, he had to present the nascent teracil backup program to Barry. Daniel was scared to death. After his presentation, Barry looked Daniel in the eyes, then looked over at the various minions around the conference table. "Good. If this doesn't work, we'll just get rid of it." Daniel thought he was talking about teracil. Alan Smith pointed out later that he was talking about the entire infectious diseases department.

It was Barry who was Brian McKinley's golfing partner and neighbor on Nantucket. It was Barry who appointed Brian as head of Penfrel Research and Development displacing the Lasker Prize winning George Finkel. And it was Barry that presided over the manufacturing fiasco that threatened Penfrel management with jail time and ultimately led to his "retirement." In the case of the manufacturing problems, it was clearly Barry who was making decisions and clearly they were his responsibility. He chose to listen to the bean counters and not to his technical experts, possibly because he understood finances and Wall Street better than the details of FDA regulations on Good Manufacturing Practice. Big mistake. But on the positive side, Barry led Penfrel to consolidate its business such that by the time he stepped down after the manufacturing scandal, Penfrel was primarily a pharmaceuticals company and no longer sold pots and pans and men's aftershave. Barry, like most large corporate CEOs, left Penfrel a very rich man.

Hank Wallace took over the reins at Penfrel after Barry's departure. He had come up through the ranks at Penfrel on the marketing side of the business. He was in the US sales division for many years. Hank, like many in the sales side of the business, was much more personable than Barry – although that says little since Barry was about as approachable as a US battle cruiser. But Hank seemed more shy and withdrawn than the irascible Barry. Hank was the one who hired Yosemite Sam to run his precious pharmaceutical research and development organization. It was Hank who shook Daniel's hand after Daniel's strategic review of infectious diseases research saying that Penfrel was behind the strategy all the way. That commitment lasted about five months when Hank presided over the near complete dissolution of infectious disease research at Penfrel. As is the case for many CEOs, next quarter's numbers and the Wall Street analysts dominated his thoughts – not products from R&D that might be years away from the market. R&D for many CEOs was just another cost center where risks were always balanced against the bottom line.

Roman at Virnuc was the founder of the company. He was an academic researcher who wanted to bring his discoveries to market to benefit himself, his friends, and mankind, perhaps in that order. He truly believed that these potential products could make a difference for patients infected with various viruses and Daniel was persuaded that he could be right. Roman was passionate. He was the complete opposite of Hank Wallace and Barry Boswell. Virnuc's potential products were Roman's personal discoveries (actually his and his friend's). His background was in the cellular toxicology of antiviral drugs. He had a deep scientific understanding of why these drugs could make people sick even while controlling their chronic viral infections.

Roman had been a consultant for virtually every company researching therapies for HIV/AIDS. He was very highly regarded by all – especially by himself. Virnuc was his company. When Daniel arrived at Virnuc and met Matt Boden, Virnuc's chief medical officer for the first time, those were almost the first words out of Matt's mouth. "Virnuc is Roman's company." Daniel came to understand that what this meant was that Roman would do what Roman wanted to do. As an executive officer at Virnuc hired for your expertise, you could only advise and then accept Roman's pronouncements calmly. This was a bit foreign for Daniel. First, he had never participated in decision-making at the CEO level before. That was because Penfrel was an organization of over 50,000 people while Virnuc was less than 100. He also never had to participate in decision-making where he came to believe that his participation counted for so little.

Roman was a shrewd and tough negotiator. He was able to stand up to executives in large pharma including the CEO of Gorman. He was always proud to say that Gorman's majority purchase of Virnuc was a deal negotiated CEO to CEO. Where this negotiating talent came from, Daniel never knew. In retrospect, Roman was Daniel's first (but not last) personal and up-front encounter with the kind of ego that could found and lead a small company trying to be independent and to bring products to a large market of patients who needed them. Even among people of that ilk that Daniel would work with as a consultant, Roman still stands out as the sine-qua-non of so-called founder's syndrome. Ego, passion, dedication, and an uncanny facility for self-delusion in the face of contrary advice and negative clinical and scientific evidence are characteristic of this disease. One of Roman's great strengths and his great weakness was his passionate dedication to his friends. It was that plus Rico's position on the Virnuc patents on its Hepatitis C drugs that ultimately led to Daniel's departure from Virnuc. Roman turned out not to be Roy Vagelos.

Gerhardt Unger was also a scientist CEO. He became CEO and chairman of Swisbiotics because of his reputation from the sale of Calibodies where he had been a founder and CEO a number of years earlier. Even though Gerhardt was a successful CEO as demonstrated by the sale, he had neither the ego nor the skills to lead an antibiotics biotech like Swisbiotics. He became a CEO entirely dominated by the investors on his board and was unable to really lead the management of the company to achieve key goals. Its true that the board of Swisbiotics was more shortsighted than that of Virnuc, but Daniel always thought that with more firm leadership from the CEO and chairman, things could have turned out differently. As it was, Swisbiotics ended up in that oblivion between a product in late stage development and the marketplace. The board could not bring itself to dilute their holdings by bringing in new investors, but at the same time they refused to put up the money required to bring their key asset to the market. Gerhardt showed himself unable to lead them out of their quandary and there they stayed. All this made Daniel wonder whether Gerhardt's success at Calibodies was due to his scientific and business skills or whether it was just based on luck.

Alistair Campbell at Parabiotics was probably the best CEO that Daniel worked with. Alistair did not have the ego of Roman and was not a scientist. He had been in business development (deal making) at one of the first large biotechs in Geneva for many years and prior to that had worked at a large pharmaceutical company doing similar work. He came to Parabiotics knowing little about antibiotics and almost nothing about the peculiarities of the clinical development and scientific discovery of new antibiotics. Daniel was amazed that Parabiotics' board would hire him on this basis. In this case, Daniel quickly recognized that the board knew exactly what they were doing. Alistair was an incredibly quick learner. He worked hard at understanding the antibiotic market during his first year at Parabiotics. He then turned his attention to trying to understand both the science of antibiotic discovery and clinical development so that he could make intelligent decisions and actually lead the company to success.

Unlike Roman, Alistair listened to expert advice and weighed his options carefully. He did not have an ego that would interfere with accepting external advice. Alistair was a talented negotiator, but he approached discussions in a more understated and positive way than Roman. During Alistair's negotiations with Green Labs, Alistair frequently went to New York to dine with the Chairman of Green. They became friends in a way. Daniel was sure that the relationship that Alistair was able to build with the Chairman of Green Labs led to the spectacular licensing deal that was beyond anything Daniel would have ever predicted for an antibiotic at the stage of development of PB-100. This deal also increased Alistair's standing with his board.

Daniel, who was a member of the board, thought that the Parabiotics board was much more supportive of the company than the Swisbiotics board. In the case of Parabiotics, different individuals held the chairman and CEO roles. Alistair could concentrate on leading the company. Although he had to put in his dues by spending lots of telephone and face time with various board members, he did not really have to manage them – he could leave that to his chairman. The chairman became someone Alistair could rely on to defend management to the board and help persuade them to do the right thing. At the same time, when Alistair's goals deviated from those of the investors, the chairman was able to quickly bring Alistair back in line without destroying company management. Parabiotics made Daniel a firm believer in the separation of the positions of CEO and Chairman.

The CEO of Marion in Basel was Jean-Marie, an ex-scientist at Roche. When he founded his company after leaving Roche with a compound that Roche had decided not to develop, he built Marion on the Roche model. Was this because that was all he knew? Daniel didn't know. But this structure seemed the opposite of what biotech should be according to Daniel. Biotech should be supple, quick to decide and quick to act. Marion had all the bureaucracy Daniel associated with large pharma companies like Roche. On the other hand, Marion was a successful company even though their adventure in antibiotics has yet to pay off.

The CEO of C-Biotech was Frank Thomas. Frank was a dealmaker like Alistair of Parabiotics. Frank was affable but had a high opinion of himself. Like Roman, Frank did not always listen to the experts he hired because of their expertise. While Daniel never knew why C-Biotech had so many deals fall apart after a year or two, he always wondered if the problem wasn't at least in part, Frank. C-Biotech still exists but Frank, like Roman, is no longer there.

Pepcore was a family owned company. The family had been involved in the chemical business in Europe for generations. They kept budgets lean – too lean in Daniel's opinion. When it came to clinical development, the owners had no previous experience nor did they know much about antibiotics.

Surprisingly, the Pepcore scientists were able to identify a compound with great antibiotic activity. They hired a Chief Medical Officer, Jose, with experience in running clinical trials – but who knew little about antibiotics. Jose tended to design their trials as if they were developing a drug for cancer where trials can be smaller and can carry higher risks than is the case for antibiotics. This also fit with the owners' bent for keeping down expenses. Because neither the CMO nor the owners understood antibiotic development very well, they were constantly confused as to which advice to follow. Daniel would tell them one thing, but the academics they spoke with would tell them something completely different. In this back-and-forth, Daniel's advice frequently took the back seat. Daniel never understood why the company continued to seek his advice, and to pay for it, but almost never followed it. Pepcore's antibiotic continues to be developed and has even been licensed by a large pharma company. Maybe Daniel was wrong and Pepcore was right all along.

Epilogue

As Daniel neared the end of his career, he looked back over the preceding 30 years. During that time, he noted sadly, there had been tremendous consolidation within the pharmaceutical industry. At some point, Daniel read a paper describing the histories of six of the large pharmaceutical companies going back to 1980. The authors found that these six companies derived from mergers and acquisitions of 70 precedent companies. If extrapolated to large pharmaceutical companies in general, this would indicate a 91% consolidation over 20 years. Since then, Daniel thought, you have to add the merger of Aventis with Sanofi to form Sanofi-Aventis, and the recent purchases of Schering-Plough by Merck and of Wyeth by Pfizer. Just a few weeks ago, he realized, Merck purchased Cubist Pharmaceuticals and closed their research site outside of Boston.

Daniel wondered how many jobs this process destroyed. Of course, this also means that there are simply fewer and fewer companies around who might be doing antibiotic research even if they were so motivated. From his perch in the consulting world, Daniel also knew that there were going to be fewer dollars coming from large companies for antibiotics. So – there would be less support for academia and small companies. And therefore, there would be less interest from investors who ultimately want their portfolio companies to be acquired by large pharma companies.

He pondered the fact that we, as societies, would need a constant pipeline of new antibiotics to stay one step ahead of resistance. Where would we train scientists to be able to discover and develop the antibiotics of the future? Daniel always hoped that could be done within those companies still pursing antibiotic research. But this opportunity seemed to teeter constantly on the brink of extinction.

To counter this devastating history, the new regulations from FDA and Europe were encouraging to the industry. More rapid and less expensive pathways to the market had been put in place. The authorities seemed less hostile. In addition, there were discussions among economists and governments focusing on providing a return on investment in antibiotics for industry. In this new environment, two companies that had abandoned antibiotics research in years past suddenly decided to plunge back in. The numbers of companies and people working in the area were increasing reversing a 20-year downward trend. When Daniel carried out his count he realized that there were four to five large companies still active in antibiotic discovery research. This is progress. Daniel even took some credit for himself in this positive transformation.

Daniel also began to add up what he thought had been his key accomplishments – delivering needed antibiotics to patients and physicians. Teracil and PB-100 have both been approved and are already being marketed. Of course, success has many fathers. But Daniel felt that he had contributed in important ways to getting these two antibiotics all the way to the marketplace. He expected another success with the approval of CY-309 in the next few years.

Daniel also thought about those ex-clients that still had drugs for which he had made important contributions in various stages of development. All carried various risks, but he thought that one or two of those might still make it all the way.

At the end of his working career, Daniel kept up with developments in antibiotic research. He still helped a few academic groups, gave lectures, wrote articles and kept up his blog. But he shut down his consulting company and turned inwards – back to Sally. He wanted to dedicate the time he had left to his family without the pressures of deadlines, egos and irascible employers. He wanted to write a book.

Acknowledgements

Unlike my first book, this one took several years to complete and I have many people to thank for their help along the way. I should first express my gratitude to those in the pharmaceutical industry, in various regulatory agencies, and the members and officers of the Infectious Diseases Society of America, for inspiring this work.

I am grateful to those who reviewed early versions of the manuscript and helped me focus on where I was going including Carole Mann, Barry Eisenstein, Lynn Silver and Mitch Bruski. Amity Shlaes has been a constant source of encouragement.

Finally, I thank my wife, Jan, for her steadfast support through all the frustrations, alterations, edits and revisions over the past several years.

About the Author

David Shlaes is a physician and microbiologist who has had a thirty year career in anti-infectives spanning academia and industry. He was Professor of Medicine at Case Western Reserve University. Later, Dr. Shlaes became Vice President for Infectious Diseases, at a large pharmaceutical company. In 1998 he was the cover feature in the April issue of Business Week dedicated to antibiotics research. He also was a member of the Forum for Emerging Infections of the National Academy of Sciences for seven years. In 2002, Dr. Shlaes became Executive Vice President for Research and Development at a biotech company located in Cambridge, MA focused on the discovery and development of antivirals. In 2005, he formed a consulting company for the pharmaceutical industry (Anti-Infectives Consulting, LLC). He worked in boardrooms and in management offices for both small and large pharmaceutical companies. Dr. Shlaes often worked with investors in the evaluation of anti-infective companies. He recently retired from Anti-infectives Consulting. His first book, Antibiotics, The Perfect Storm (Springer), was published in 2010. He remains an Editor for the journal, Antimicrobial Agents and Chemotherapy, writes a blog called, "Antibiotics the Perfect Storm," and remains active in antibiotic policy areas. "The Drug Makers" is his first work of fiction. David lives with his wife in Eastern Connecticut.